G000122566

Digital Electronic Circuits and Systems

Noel M. Morris

Principal Lecturer
North Staffordshire Polytechnic

© Noel M. Morris 1974

All rights reserved. No part of this publication may be reproduced or transmitted, in any form or by any means, without permission

First published 1974
Reprinted 1977, 1978, 1980, 1982

Published by
THE MACMILLAN PRESS LTD
London and Basingstoke
Companies and representatives throughout the world

ISBN 0 333 14862 2

Printed in Hong Kong

HARROW COLLEGE OF
HIGHER EDUCATION LIBRARY

The paperback edition of this book is sold subject to the condition that it shall not, by way of trade or otherwise, be lent, re-sold, hired out, or otherwise circulated without the publisher's prior consent in any form of binding or cover other than that in which it is published and without a similar condition including this condition being imposed on the subsequent purchaser

Contents

Foreword

Technological progress has nowhere been more rapid than in the fields of electronic, electrical, and control engineering. The *Macmillan Basis Books in Electronics* have been written by authors who are specialists in these fields, and whose work enables them to bring technological developments sharply into focus.

Each book in the series deals with a single subject so that undergraduates, technicians, and mechanics alike will find information within the scope of their courses. The books have been carefully written and edited to allow each to be used for self-study; this feature makes them particularly attractive not only to readers approaching the subject for the first time, but also to mature readers wishing to update and revise their knowledge.

Noel M. Morris

Preface

Developments in digital electronics, particularly since the introduction of large-scale integrated circuits, have led to the widespread use of digital systems in almost every walk of life. This book provides coverage of aspects of digital electronics ranging from basic gates and logical algebra to sophisticated systems.

Readers can confidently begin their studies in digital systems with this book, since the studies deal both with the operating principles of logic gates and with their logical functions. Boolean algebra and Karnaugh mapping techniques are used as design tools, enabling readers to understand the design philosophy of advanced systems.

Included in the book are all the important facets of logic systems including logic families, integrated circuits, field-effect logic gates, arithmetic processes, and electronic counters. The book culminates in a chapter on applications of logic which includes many popular circuits used both in commercial and professional equipment.

The electronics industry has kindly supplied me with valuable information relating to the circuits and systems described in the book, and I would like to record my thanks for the help received. Finally, the book could never have been written without the help, patience, and understanding of my wife and family.

Noel M. Morris
Meir Heath

1 What is Logic?

1.1 Gates

The subject of electronic logic is one which embraces the whole field of electronics from computers to automobiles and from telephone exchanges to toys. 'Logic' devices serve man in every walk of life, each device or system operating in a predictable manner. In fact, so predictable is their operation that we can use a form of **logical algebra** to determine the way in which the circuit works. This type of algebra is sometimes known as **boolean algebra,** after the Rev. G. Boole (1815–64) who set down the basic rules. Mathematicians often refer to it as **set theory,** or the theory of 'sets'.

The basic rules of this form of algebra are quite simple and, once understood, are relatively simple to apply. Devices used in logic networks control the flow of *information* through the system and, for this reason, are known as *logic gates* since the 'gates' are opened and closed by the sequence of events occurring at their inputs. The basic range of gates are known by the names **AND, OR, NOT, NOR,** and **NAND,** and are described in detail in the chapters which follow.

The operation of each gate or system is defined by a logical algebraic statement, the logical equation being amenable to manipulation by the rules of boolean algebra. Thus we see that boolean algebra is a means of setting down the operation of logic circuits and networks in the form of a series of 'equations'.

Many applications of logic networks, such as those involved in counting and other arithmetic processes, require the use of **MEMORY** elements which retain or store information. Memory elements are constructed simply by interconnecting a number of the basic gates mentioned above in such a way that the circuit retains the original input data after the input signal has been removed. The information stored by the memory can be changed or updated by subsequent control signals.

1.2 Logic signal levels

In the world of logical algebra every question has a definite solution, so that all problems are given a 'yes' or 'no' type of solution, that is, the solution is either true or false. Thus, we are dealing with a binary or two-level system.

Digital electronic circuits have specific bands of voltage levels allocated to them which conform to the 'true' and 'false' conditions. If, for example, we have an electronic switch whose supply voltage is +5 V then, depending on the external load connected, the 'true' output condition may be represented by any voltage in the range +3 to +5 V, and the 'false' output may be represented by a voltage in the range zero to +0.5 V. Using a notation known as positive logic, the 3–5 V range is described as the logic '1' level, and the 0–0.5 V range is described as the logic '0' level. In this notation, logic '1' is a 'true' solution, and logic '0' is a 'false' solution.

Another notation which is sometimes used is the negative logic notation, in which the 3–5 V range is described as logic '0', and the 0–0.5 V range is described as logic '1'.

The positive logic notation is the one most frequently used in connection with logic circuits, although there are exceptions to this, which are mentioned as they arise in the text.

2 Basic Logic Functions

In order to outline the operation of the logic devices mentioned in chapter 1, let us consider the operation of a coin circuit of a hypothetical coin-operated vending machine.

Suppose that our machine dispenses a beverage after either a 10p or a 5p coin has been inserted into a coin slot. The cost of our drink is to be 5p, so that when a 10p coin is inserted, the machine returns 5p to us. If a 5p 'change' coin is not available in the machine, a sign stating 'USE CORRECT CHANGE ONLY' is to be illuminated. In the following sections we will consider how the functions of the coin circuit may be carried out by logic gates.

2.1 The AND function

In this section we shall consider the circuit associated with the release of the 5p change. Prior to inserting a 10p coin we must take note that the 'USE CORRECT CHANGE ONLY' light is extinguished, which implies that the machine holds at least one 5p coin. Thus, when the 10p coin is inserted AND the 'CORRECT CHANGE' light is extinguished, then the coin release circuit is activated. If we assign symbol T to the output of the detector which senses the 10p piece, symbol C to the output of the sensor associated with the 5p coin stored in the machine, and symbol X to the output of the logic gate which initiates the coin release mechanism, then

$$X = T \text{ AND } C = T \, . \, C$$

The 'dot' (.) symbol is used in logical equations to represent the logic AND function.

Sensor T provides a logic count '1' output signal when a 10p coin is inserted, and a logic '0' when no 10p coin is present. Similarly, the output from sensor C is '1' when the machine holds a 5p coin, and a '0' when it does not. Thus $X = 1$ (that is, a 5p coin is released) when $T = 1$ and $C = 1$. If either $T = 0$ or $C = 0$, then $X = 0$ and the operation of the coin release circuit is inhibited.

If we collect together all the possible operating combinations of the AND gate associated with the coin release circuit, we have what is known as a *truth table* which, for the two input AND gate, is given in table 2.1.

A relay circuit which satisfies the AND function is shown in figure 2.1. Here the signals T and C are derived from the sensors described above, and output X is used

Table 2.1
Truth table for a two-input AND gate

inputs T	C	output $X = T \, . \, C$
0	0	0
0	1	0
1	0	0
1	1	1

to energise the coin release circuit. In our relay circuit, when either signal T or signal C is logic '0', the appropriate relay contacts are open and the output voltage from the circuit is zero. When $T = C = 1$, both contacts are closed, and the full voltage (logic '1') appears at the output. A number of symbols are used to represent the AND gate, two popular versions being shown in the figure.

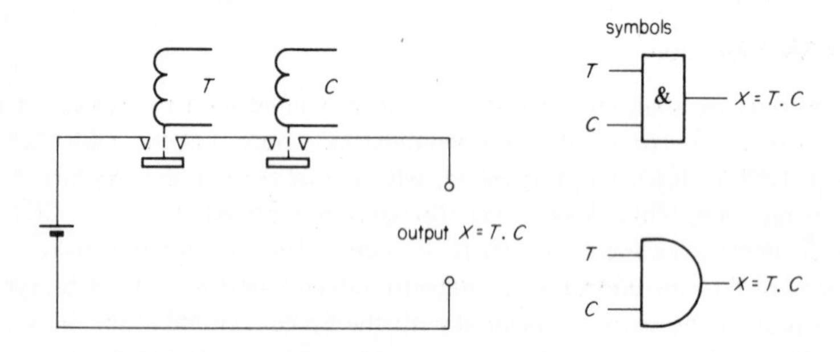

Fig. 2.1 A logic AND gate

In a complex system the AND gate may have a large number of inputs, say N inputs, and if the output from the circuit is f, then the logical equation describing the operation of the gate is

$$f = A \, . \, B \, . \, C \, . \, \ldots \, L \, . \, M \, . \, N$$

This corresponds to a relay circuit with N relays in series, each with a normally open pair of contacts, and each relay being energised by a separate input signal.

2.2 The OR function

In the specification of our vending machine we said that the drink was prepared if either a 10p coin or a 5p coin is inserted into the coin slot. If we assign the symbol F to the sensor which detects 5p pieces and, as before, sensor T detects 10p coins, and we let symbol Y represent the output from the gate, then

$$Y = F \text{ OR } T = F + T$$

The 'plus' (+) symbol is used throughout this book to represent the logical OR function. This symbol should not be confused with the arithmetic addition symbol, the difference between the two symbols being explained later in this section. An alternative symbol sometimes used for the OR function is a 'vee' (v), that is, $Y = F \vee T$. The truth table for the OR function described above is given in table 2.2.

Table 2.2
Truth table for a two-input OR gate

inputs F	T	output $Y = F + T$
0	0	0
0	1	1
1	0	1
1	1	1

From the truth table we see that the output from the gate is '1' when either input signal is '1'. An interesting situation arises if we insert a 5p and a 10p coin simultaneously (assuming, of course, that this can be done!), since the machine accepts both coins and dispenses only one drink! This condition is illustrated in the final line of the truth table, in which the output signal from the gate is '1' when both inputs are activated. Thus the logical statement that $1 + 1 = 1$ is valid; it should *not* be confused with the arithmetic addition operation.

Also, from the truth table, the output from the gate is zero when both inputs are zero, that is, when no coins are inserted.

A logic '1' output from the OR gate causes the vending circuits to be activated, so producing our drink.

A relay circuit which satisfies the OR function is shown in figure 2.2. In this case the circuit provides an output voltage when either F or T or both input signals are

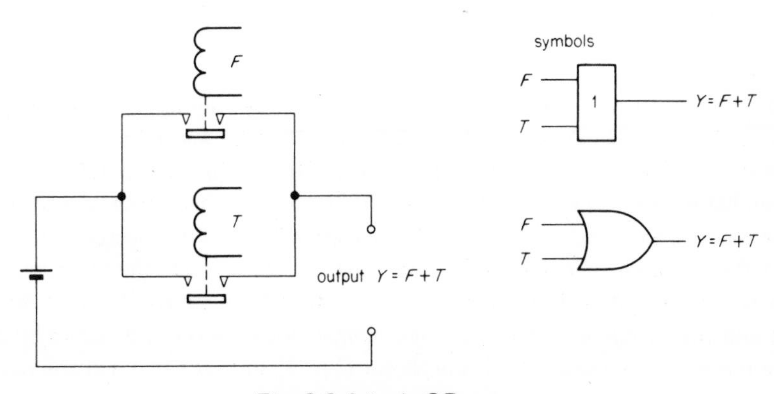

Fig. 2.2 A logic OR gate

present. This circuit is sometimes known as an **inclusive**–OR gate, since it provides an output signal in the case when F and T are both '1'. Later in the book we shall deal with yet another type of gate known as an **exclusive**–OR gate, which gives an output of logic '0' when both F and T are energised by logic '1' signals.

Some circuits require the use of multiple-input OR gates, and if a large number of inputs, say N inputs, are employed then the logical output f developed by the gate is

$$f = A + B + C + \ldots + L + M + N$$

This corresponds to a relay circuit with N relays in parallel, each having a normally open pair of contacts, and each relay being energised by a separate signal.

2.3 The NOT function

When discussing the signal levels associated with logic circuits, we saw that only two steady values could exist, that is, logic '1' and logic '0'.

When the output from a logic gate is '1', quite clearly it is not '0'. Also, when the output is '0' it is not '1'. Thus a gate which generates the NOT function provides an output of logic '0' when the input is '1' and vice versa. The process of logically inverting or complementing (that is, the NOT function) a function is signified by placing a 'bar' over the function, as shown below for the function C.

$$\text{NOT } C = \overline{C}$$

The truth table for this function is given in table 2.3.

Table 2.3
Truth table for a NOT gate

input C	output $\overline{C}$
0	1
1	0

Let us consider how a NOT gate can be used in the coin circuit of our vending machine. In the original specification of the machine we said that if a 5p piece is not available in the machine, then the 'USE CORRECT CHANGE ONLY' light is to be illuminated. Hence, we can illuminate the lamp from the output of a NOT gate whose input is energised from sensor C. Readers will recall that sensor C is used to detect the presence of coins held in the 'change' stack inside the machine. Thus when no change is held in the machine, the output from sensor C is zero, so that $\overline{C} = 1$ and the change warning lamp is illuminated. When sensor C detects a 5p piece in the machine, the output from the sensor C is '1' so that $\overline{C} = 0$ and the warning lamp is extinguished.

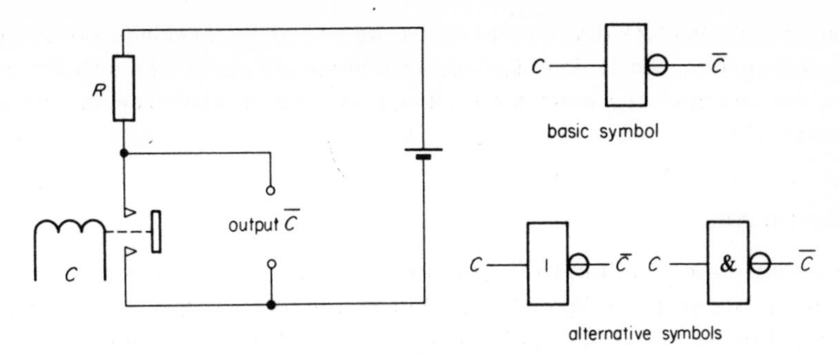

Fig. 2.3 A logic NOT gate.

One form of relay NOT gate is illustrated in figure 2.3. When $C = 0$, the relay is de-energised and the output voltage is HIGH, that is, $\overline{C} = 1$. When $C = 1$, the relay is energised and the output terminals are short-circuited together so that $\overline{C} = 0$. Resistor R is included in the circuit to limit the current drawn from the supply when the relay contacts are closed.

2.4 Complete coin circuit of the vending machine

A block diagram of the logic circuit of the coin section of our vending machine is shown in figure 2.4. Inputs F and T are activated by placing a coin in the appropriate slot in the machine, and signal C is generated by the presence of a 5p coin inside the machine.

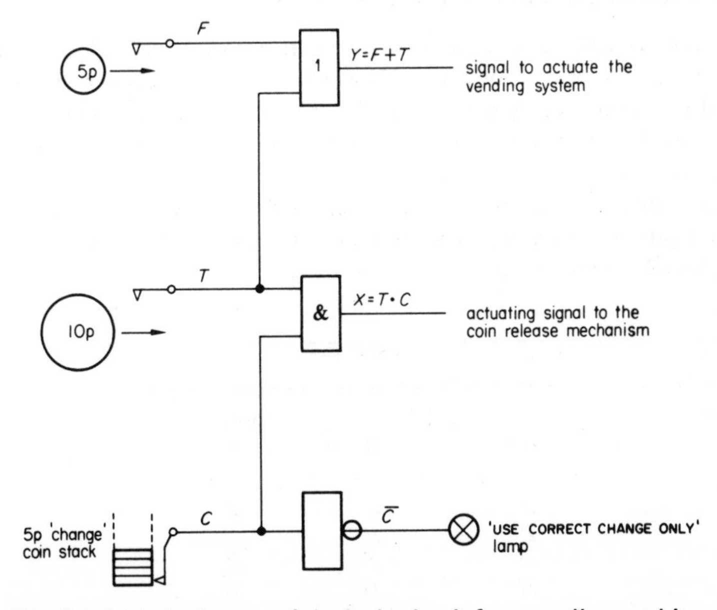

Fig. 2.4 A block diagram of the logic circuit for a vending machine

When using the block diagram technique, we only show the connections through which information signals flow. Connections which are concerned with the power supply, for example, the main supply line, bias supplies, earth line, et cetera, are not shown.

2.5 Negated inputs

Some circuits have a built-in NOT gate associated with individual inputs, shown in the case of input C in figure 2.5(a). Where this occurs, the negated input is represented by a circle at the input, shown in figure 2.5(b). A circuit using this type of symbol is described in chapter 13.

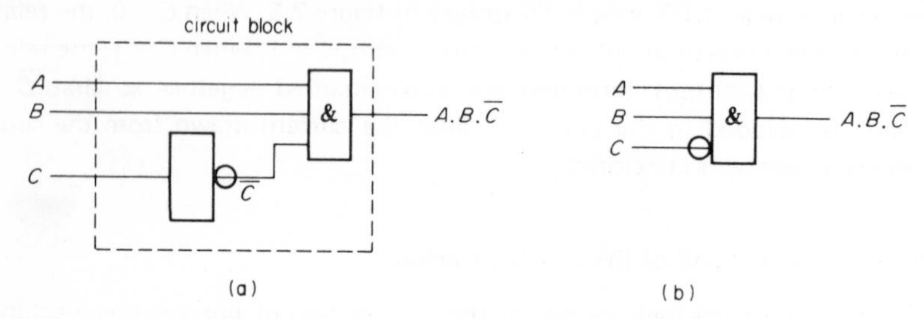

Fig. 2.5 A gate with a negated (INVERTING) input

2.6 The Effect of positive and negative logic levels

In the first chapter it was pointed out that a specific electrical voltage may represent either a logic '1' or a logic '0' signal. So far in the book we have talked in terms of a positive voltage representing '1', and zero potential representing '0'. Let us now consider the effect of operating a given logic gate alternately with positive and negative logic levels.

Suppose that we have measured the input and output voltages associated with a two-input gate, and have found them to be as shown in table 2.4, where H is a high voltage, and Z is zero voltage.

Table 2.4

inputs A	inputs B	output X
Z	Z	Z
Z	H	Z
H	Z	Z
H	H	H

Positive logic operation

In the positive logic notation, $H = 1$ and $Z = 0$. If we rewrite the voltage levels in table 2.4 in terms of positive logic levels, we obtain table 2.5.

Table 2.5

inputs A	B	output X
0	0	0
0	1	0
1	0	0
1	1	1

Comparing table 2.5 with the truth table for a two-input AND gate, table 2.1, we see that when we use the positive logic notation the gate generates the **AND function** of the inputs.

Negative logic operation

Using the negative logic notation, $H = 0$ and $Z = 1$. Rewriting the voltage levels in table 2.4 in terms of negative logic, we obtain the results in table 2.6.

Table 2.6

inputs A	B	outputs X
1	1	1
1	0	1
0	1	1
0	0	0

Let us now compare table 2.6 with table 2.2 for a two-input OR gate. Comparing **like** input conditions in both cases, we see that table 2.6 is that of an OR gate.

Summary

Clearly the name we give to a logic gate (defined by its truth table) depends on the logic notation used in the system. As we have seen above, a gate which is described as an AND gate when using the positive logic notation also operates as an OR gate when using negative logic notation. The reader may also like to show that a positive logic OR gate generates the negative logic AND function.

Positive logic devices and systems will largely be considered in the remainder of the book. Specific reference will be made to circuits using negative logic.

3 *NAND and NOR Functions*

The names NAND and NOR are contractions of the following logic functions

$$\text{NAND} = \text{NOT AND} = \overline{\text{AND}}$$

$$\text{NOR} = \text{NOT OR} = \overline{\text{OR}}$$

The functions are described in detail in the following sections.

3.1 The NAND function

The NAND function is generated when an AND gate and a NOT gate are combined in the manner shown in figure 3.1, and its truth table is given in table 3.1.

Table 3.1
Truth table for a two-input NAND gate

inputs		intermediate AND function $A \,.\, B$	output $\overline{A \,.\, B}$
0	0	0	1
0	1	0	1
1	0	0	1
1	1	1	0

We shall consider the operation of the circuit in terms of the two stages in figure 3.1. The first stage of the circuit generates the AND function which, as we saw earlier, gives an output of '0' whenever any input is '0', and an output of '1' only when both inputs are '1', resulting in the intermediate AND result in table 3.1. The NOT section of the NAND gate complements or inverts the intermediate AND signal to give the final output. The complete truth table can be summarised as follows:

When any input to a NAND gate is energised by a logic '0' signal, then its output is '1'. Otherwise the output is '0'.

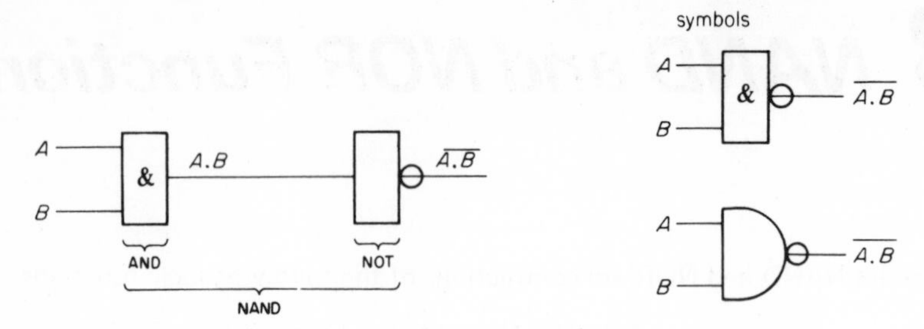

Fig. 3.1 A circuit which generates the NAND function

3.2 The NOR function

A block diagram of a network which generates the NOR function is shown in figure 3.2, having the truth table in table 3.2.

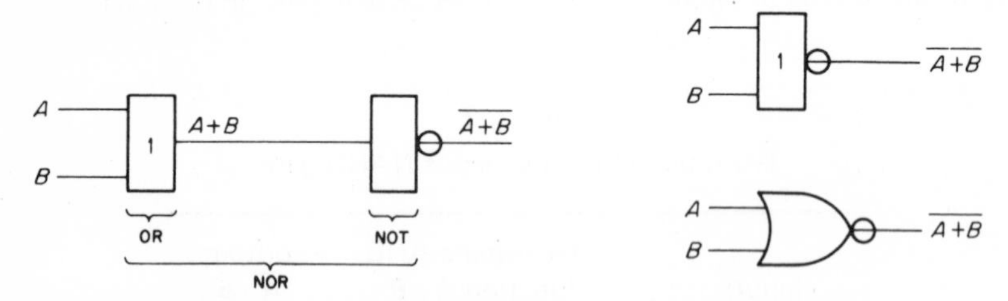

Fig. 3.2 A circuit which generates the NOR function

Table 3.2
Truth table for a two-input NOR gate

inputs		intermediate OR function $A + B$	output $\overline{A + B}$
0	0	0	1
0	1	1	0
1	0	1	0
1	1	1	0

The OR section of the circuit generates a logic '1' whenever a '1' signal is applied to either input, and this is inverted by the NOT gate. The overall truth table may be summarised as follows:

When any input to a NOR gate is energised by a logic '1' signal, then its output is '0'. Otherwise the output is '1'.

3.3 Why use NAND and NOR gates?

A feature of both NOR and NAND networks is that they can be used to generate the truth tables of all other types of gates; that is, a number of NAND gates connected in particular configurations can be used to generate the AND, OR, NOT, and NOR functions (see chapter 7), and various types of memory circuits can also be constructed (see chapter 8). This has obvious economic advantages in so far as it is necessary for users to purchase and to stock only one basic type of logic element.

At first sight it may seem that a large number of either NAND or NOR gates are required to replace an equivalent AND-OR-NOT type of network. This, in fact, is not necessarily the case since boolean algebraic techniques can be used to **minimise** the number of gates required in the system. In many instances it is possible to generate a given logic function using a smaller number of either NAND or NOR gates than is possible with AND, OR, and NOT gates. Moreover, the cost of NAND and NOR elements is generally less than that of other types of elements.

3.4 The effect of positive and negative logic conventions on NAND and NOR gates

We can illustrate the effect of employing either positive or negative logic conventions on the truth tables of these elements by considering table 3.3.

Table 3.3

inputs		output
A	B	
H	H	Z
H	Z	Z
Z	H	Z
Z	Z	H

This is a table of input and output voltages of a particular logic gate, where H is a high voltage and Z is zero voltage. In positive logic $H = 1$ and $Z = 0$, and in negative logic $H = 0$ and $Z = 1$. Applying these relationships to table 3.3 we obtain tables 3.4 and 3.5 for positive logic operation and negative logic operation, respectively.

Table 3.4
Positive logic truth table

inputs		output
A	B	
1	1	0
1	0	0
0	1	0
0	0	1

Table 3.5
Negative logic truth table

inputs		output
A	B	
0	0	1
0	1	1
1	0	1
1	1	0

Comparing table 3.4 with the truth table for a 2-input NOR element, table 3.2, we see that the device may be regarded as a **positive logic NOR gate.** Comparing table 3.5 with the truth table for a 2-input NAND element, table 3.1, we also see that it may be regarded as a **negative logic NAND gate.**

It is for this reason that it is sometimes necessary to state whether positive or negative logic conventions are being used when specifying the name of the logic function generated by the gate. In some cases the gate is described as a NAND/NOR gate, and the manufacturer provides the truth table which indicates the voltage levels involved, that is, a truth table similar to table 3.3.

4 Electronic Switches

An ideal switch is one which, when it is 'on' has no effective resistance between its terminals and no voltage appears across it when current flows through it. When it is 'off', it has an infinite resistance between its terminals, and the leakage current through it is zero. Moreover, our ideal switch has a perfect switching action so that the time taken for the circuit current either to be cut off or to be switched on is zero.

No practical device can hope to attain these ideals, but it is the semiconductor diode and the transistor which most nearly approach them.

4.1 Semiconductor materials

Transistors and diodes are manufactured from a range of materials known as **semiconductors,** and have certain valuable properties which allow them to be used to control the flow of current in electronic circuits. A semiconductor is a material whose conductance at room temperature is mid-way between that of a good conductor and that of a good insulator. In quantitative terms, the conductance of semiconductors lies in the range $10^{-4}\,\Omega$m to $10^{3}\,\Omega$m.

The two principal semiconductor materials used for the construction of transistors and diodes are **silicon** (Si) and **germanium** (Ge), the former being universally used in electronic logic devices. In the early years of semiconductor electronics, germanium was widely used since the production technology associated with it was well understood. Silicon has a number of advantages in switching applications over germanium, including a wider operating temperature range and a lower leakage current. Silicon production techniques have overtaken those of germanium, and present generations of silicon devices have superior performance parameters than those of germanium devices. Other types of semiconductor materials, for example, gallium arsenide (GaAs), are used in specialised applications and are not of interest to us here.

Semiconductors can exist in three basic forms, which are ***i*–type, *p*–type,** and ***n*–type. Intrinsic** semiconductors (*i*–type) are commercially pure semiconductors. The conductivity of this type of material increases with temperature; that is, it has a **negative** resistance-temperature coefficient, and the flow of current through it is largely temperature dependent. Ideally, *i*–type semiconductors should be good insulators at room temperature, silicon approaching this more closely than germanium.

Electrons in the **outermost** orbit of atoms are known as **valence electrons,** and are responsible for the chemical and electrical properties of the atom. Silicon and germanium atoms have four valence electrons, and are known as **tetravalent** atoms. Electrical conduction occurs as a result of the movement in an electric field of valence electrons which have become detached from parent atoms. Since the negative charge of the valence electrons (− 4 units of charge for a tetravalent atom) just balances the positive charge (+4 units) on the remainder of the atom, the **net** electrical charge on an isolated atom is zero. When an electron leaves the atom, it takes with it a negative charge of − 1 unit and leaves behind an atom with a **net** positive charge of +1 unit on it. This positive charge is described as a **hole,** and is equivalent to an electron deficiency. Thus current flow is due to both the flow of electrons towards the positive pole of the supply and of holes towards the negative pole. The type of charge carrier primarily involved in the conduction process depends on the type of semiconductor in use.

In *i*–type semiconductors, the valence electrons are all required for chemical bonding purposes, and for this reason *i*–type materials are poor conductors.

If we deliberately introduce a known amount (about 1 part in 10^8) of a specific type of impurity, the electrical properties of the semiconductor are altered in a particular way. If, for example, we introduce an impurity atom which has five valence electrons **(pentavalent atoms)** into silicon or germanium, only four of the five valence electrons are required for chemical bonding purposes. The remaining electron can easily be detached for conduction purposes, so that this electron becomes a **mobile charge carrier** within the atomic structure. Since the mobile charge carrier is an electron it has a negative charge on it, and the material is known as an ***n*–type** semiconductor. In *n*–type materials, since electrons are mobile, *current flow is due primarily to the movement of negative charge carriers from the negative pole of the supply to the positive pole.*

Pentavalent impurity substances include arsenic (As), phosphorus (P), and antimony (Sb).

On the other hand, if we introduce an impurity atom which has three electrons in its outer orbit **(trivalent atoms),** such as aluminium (Al), boron (B), gallium (Ga), or indium (In), the resulting material has an electron deficiency in its structure. That is, *conduction takes place as a result of the movement in an electric field of mobile positive charge carriers or holes.* This type of material is known as a ***p*–type semiconductor.**

4.2 Semiconductor junction diodes

The simplest type of electronic switch is the semiconductor diode, shown in figure 4.1. This device is constructed in a single crystal of semiconductor which contains a *p*–region and an *n*–region. The *p*–region is the **anode** of the diode, and the *n*–region is its **cathode,** and when the anode is positive (*p–region positive*) with respect to the *n*–region, the diode is said to be **forward biased** and current flows through it. Under these conditions the **forward voltage drop** across the diode when

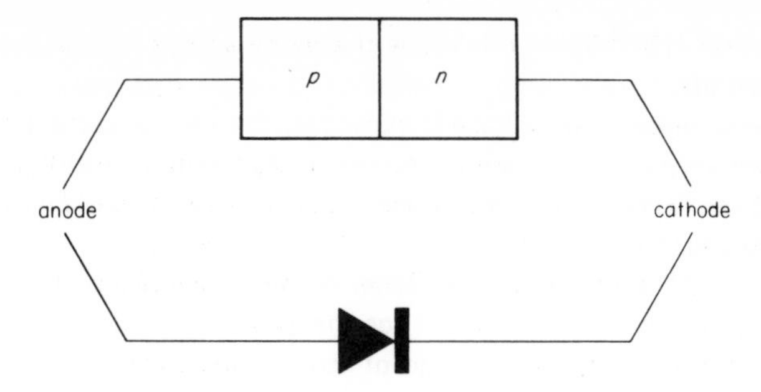

Fig. 4.1 A *p–n* junction diode

carrying its rated current is about 0.7–0.8 V in the case of a silicon device, and about 0.3–0.4 V in the case of a germanium diode. The diode is said to be in its **forward conduction mode,** corresponding to operation in the first quadrant of the characteristic in figure 4.2. In this mode, the diode operates as a switch which is turned **on.**

If we connect the anode to the negative pole of the supply and the cathode to the positive pole, the diode cuts off the flow of current and is then said to be in its **reverse blocking mode.** Providing that the **reverse voltage** or **inverse voltage** applied to the diode does not exceed its **breakdown voltage,** only **leakage current** flows between the two regions. The leakage current in silicon devices may be as low as a

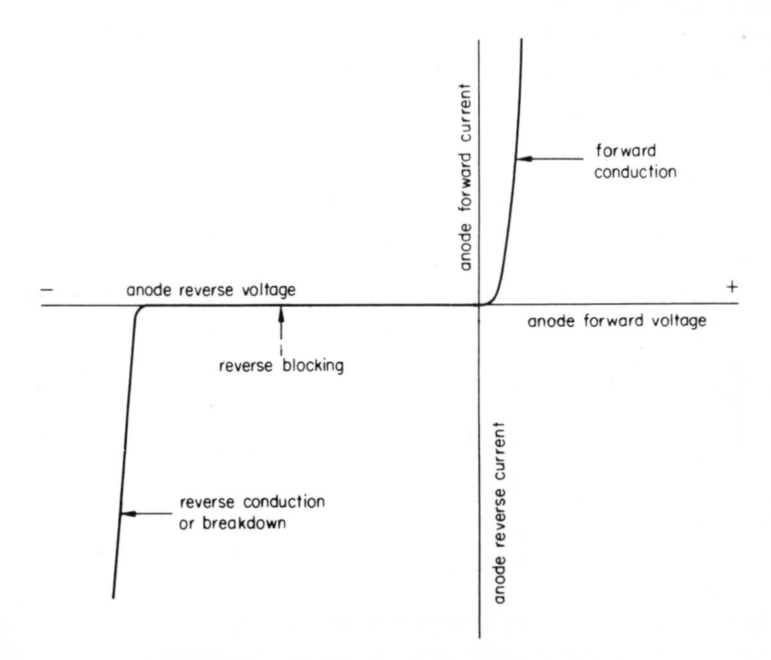

Fig. 4.2 Characteristic of a *p–n* junction diode

few nanoamperes (10^{-9}A). In this mode the diode acts as though it were a switch which is turned **off.**

If the reverse breakdown voltage is exceeded, the reverse current increases at a rapid rate with very little increase in reverse voltage. With conventional diodes this operating region is avoided, since it can rapidly lead to excessive heating with consequent damage to the diode.

Certain types of diode, known as **Zener diodes,** are designed to operate in the reverse breakdown mode. The symbol for the Zener diode is shown in figure 4.3. Zener diodes are used in a wide range of circuits including bias circuits, voltage offset circuits, voltage reference circuits, and stabilised power supplies.

Fig. 4.3 Circuit symbol of the Zener diode

4.3 Charge carrier storage in diodes

When the bias applied to a junction diode changes from forward bias to reverse bias, the charge carriers within the device must recombine with parent atoms and disappear before the current through the diode can fall to zero. This action results in a pulse of reverse current flowing through the diode, in the manner shown in figure 4.4. Here the circuit is limited in magnitude to V/R by the circuit resistance. The effect which causes the reverse current is known as **charge storage**; the net effect of charge storage is to introduce a delay between the time that the supply

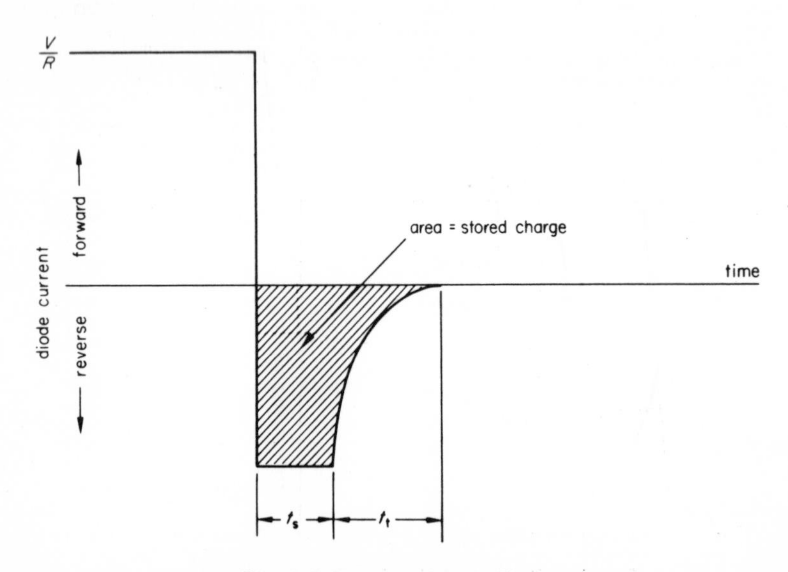

Fig. 4.4 Charge storage effect

voltage is reversed and the time that the anode current falls to zero (or, more accurately, to its leakage value).

The **storage time** t_s of the diode is the time taken for the reverse voltage to sweep the charge carriers away from the junction; the **transition time** t_t is the time taken for the reverse current to fall to zero. The total delay is sometimes quoted as the **recovery time** of the diode and, in switching diodes, can be as small as a few nanoseconds (10^{-9}s). Diodes used in switching circuits should have as small as possible a recovery time.

Charge storage has a limiting effect on the ultimate switching speed of all semiconductor devices employing *p–n* junctions whose bias is reversed from time-to-time. The bipolar junction transistor (see section 4.5) is no exception to this.

4.4 The Schottky barrier diode

The Schottky barrier diode is a device with a metal-to-semiconductor (usually *n*–type) rectifying junction. The operating principle of this device differs from that of the *p–n* junction diode, and it does not exhibit the storage time delay associated with it. The latter factor makes the Schottky barrier diode an attractive device for some applications in switching circuits (see also section 4.13).

4.5 The bipolar junction transistor (BJT)

The bipolar junction transistor is a device constructed in a single crystal of semiconductor material, and has three regions known respectively as the **emitter,** the **base,** and the **collector.** Figure 4.5 shows a pictorial representation of the two

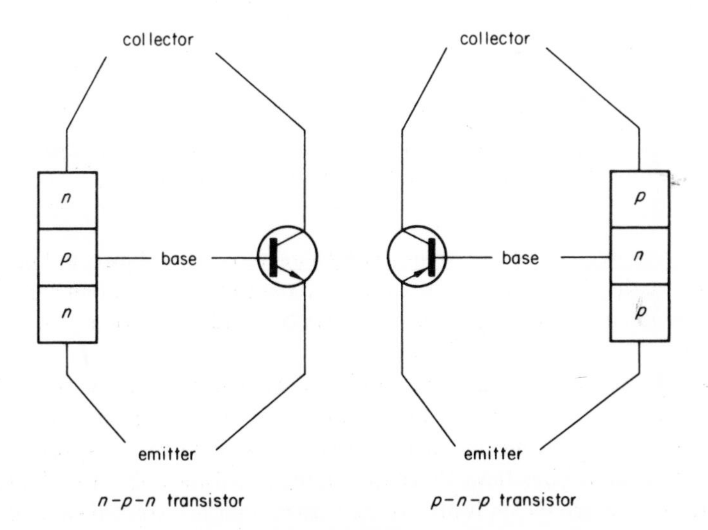

Fig. 4.5 Circuit symbols for bipolar junction transistors

main types of bipolar transistor, known as *n–p–n* and *p–n–p* transistors respectively. During the process of conducting current, both types of BJT employ both holes and electrons in their operation, and it is for this reason that they are described as *bipolar* devices.

In both types of transistor the emitter is the region which emits charge carriers into the transistor, the base is the region used to control the flow of current through the device, and the collector is the region in which the charge carriers are collected.

The manner in which the base current controls the collector current is best illustrated by means of the **output characteristic** or **collector characteristic** of the transistor, an example of which is shown in figure 4.6(a). The characteristics shown are the **common-emitter characteristics** of an *n–p–n* transistor, and are obtained by testing the transistor with the emitter region used as the connection that is common to both the base signal (that is, the control signal) and the collector supply voltage.

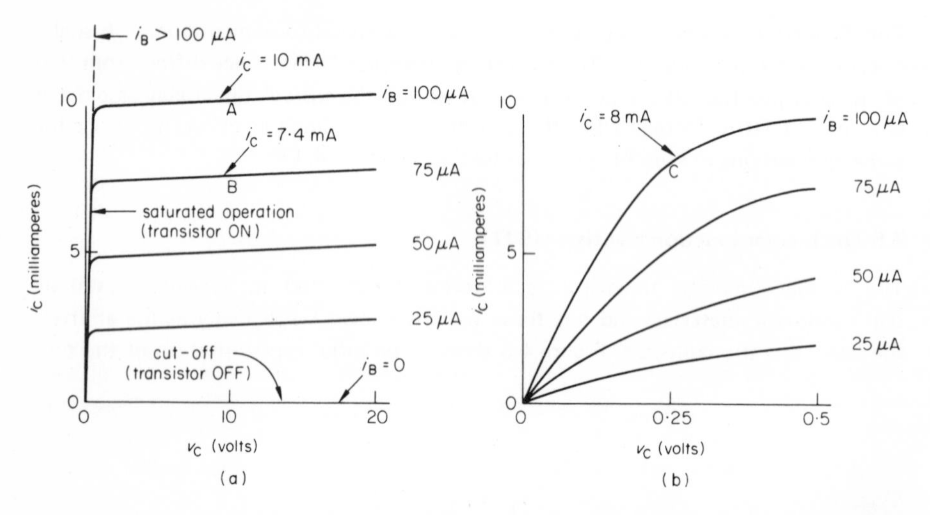

Fig. 4.6 (a) Common-emitter collector characteristics for an *n–p–n* transistor and (b) the collector characteristic in the region of the origin

A typical test circuit used to obtain these characteristics is shown in figure 4.7. In switching circuits, BJT's are almost invariably used in this operating mode (known as the *common-emitter mode*) since it offers both a high current gain and a high power gain.

We see from figure 4.6(a) that when the base current i_B is zero, then the collector current is also zero. In fact this is not quite true, since a small value of leakage current (usually in the range between a few nanoamperes (nA) and a few microamperes (μA)) does flow. When $i_B = 0$, the transistor is said to be **cut off,** and it operates as a switch which is **off.** As the base current is increased we find that the collector current also increases. With a base current of 75 μA we see that the

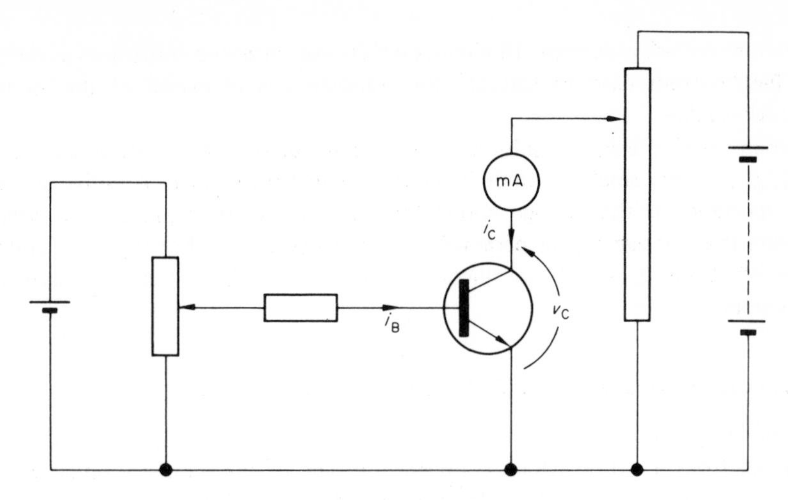

Fig. 4.7 A circuit used to determine the output characteristic

collector current is 7.4 mA, and a base current of 100 μA gives a collector current of 10 mA.

An important parameter of the BJT is its **forward current gain,** which is defined under a variety of conditions discussed below. The absolute current gain or **d.c. forward current gain,** designated the symbol h_{FE}, is the ratio of the total collector current to the total base current. At the point A on figure 4.6(a)

$$h_{FE} = 10\ \text{mA}/100\ \mu\text{A} = 100$$

This parameter is of particular significance when bias conditions are being evaluated. The **small signal forward current gain** h_{fe} is the ratio of the **change** occurring in the collector current when the base current is changed in value; this ratio is determined at a constant value of collector voltage (10 V in figure 4.6(a)). The value of h_{fe} determined between the points A and B on figure 4.6(a) is

$$h_{fe} = (10 - 7.4)\ \text{mA}/(100 - 75)\ \mu\text{A} = 104$$

Parameter h_{fe} is of value in the design of small signal linear amplifiers.

Of particular importance in switching circuit design is the **saturated forward current gain,** designated the symbol $h_{FE(sat)}$. A BJT is said to be saturated when its operating point lies on the steep part of the curves on the extreme left-hand of figure 4.6(a). When in the saturated state, the transistor carries a large current and supports only a small voltage across it. This section of the characteristics is expanded in figure 4.6(b). The value of $h_{FE(sat)}$ evaluated at point C at a collector voltage of 0.25 V is

$$h_{FE(sat)} = 8\ \text{mA}/100\ \mu\text{A} = 80$$

Thus, to drive the transistor into saturation, that is, to cause it to operate as a switch in the ON condition, the base current must be **at least** 1/80th of the maximum collector current. In circuits which use saturated transistors as switches, the base current used to saturate the transistors is in excess of the minimum saturation value.

When the transistor saturates, the **collector–emitter saturation voltage,** $V_{CE(sat)}$, is very small, its value being 0.25 V in figure 4.6(b). Another parameter of importance is the voltage applied to the base–emitter junction in order to saturate the transistor. This parameter is designated the symbol $V_{BE(sat)}$, and has a value of about 0.7–0.8 V in silicon transistors and about 0.4 V in germanium transistors.

4.6 A resistor–transistor logic (RTL) NOT gate

A basic form of RTL NOT gate is shown in figure 4.8. Input A applied to the gate is obtained from a switch which connects the base of the transistor either to earth (logic '0') or to positive voltage V (logic '1'). When $A = 0$, as shown in the diagram, the base current is zero and the transistor is cut off. In this case, the collector current is zero and the output signal ($\overline{A}$) is logic '1'. If the output is unloaded, then $I_L = 0$ and the output voltage is equal to V_{CC}. When the output is loaded, the current flowing in R_L causes the output voltage to fall below V_{CC}, and it is not unusual to design circuits to operate with a logic '1' level in the range 0.5 V_{CC} to V_{CC}.

When input A is switched to V_1, that is, to logic '1', the current flowing in R_1 drives the transistor into saturation, so that output $\overline{A}$ is logic '0'.

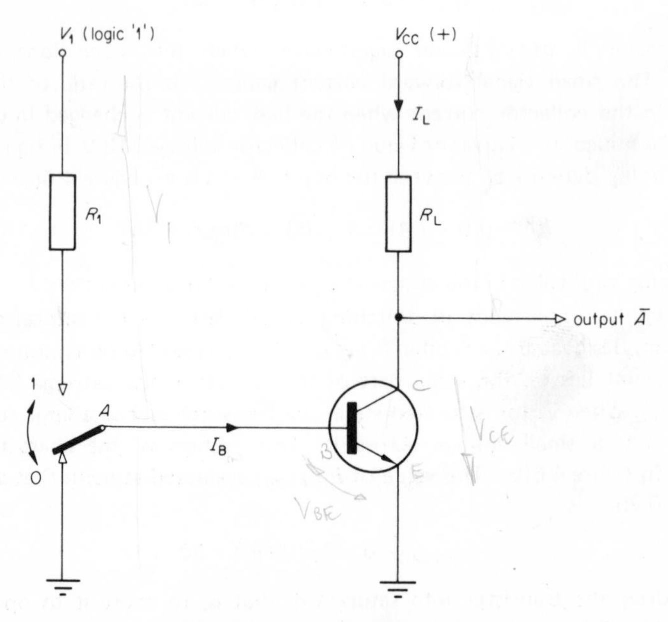

Fig. 4.8 An RTL NOT gate

We will now consider the design of an RTL NOT gate, the design procedure producing a circuit which will operate satisfactorily in an *unloaded* state. The design is then extended to allow for the effects of loading and other factors.

When the transistor is saturated, the current $I_{L(sat)}$ flowing in the load is

$$I_{L(sat)} = \frac{V_{CC} - V_{CE(sat)}}{R_L} \approx \frac{V_{CC}}{R_L}$$

The simplification in the above relationship is brought about by the assumption that $V_{CE(sat)}$ is much smaller than V_{CC}. The base current $I_{B(sat)}$ needed to saturate the transistor is

$$I_{B(sat)} = \frac{I_{C(sat)}}{h_{FE(sat)}} = \frac{V_{CC}/R_L}{h_{FE(sat)}}$$

where $I_{C(sat)}$ is the saturation current flowing through the transistor. Now

$$I_{B(sat)} = \frac{V_1 - V_{BE(sat)}}{R_1} \approx \frac{V_1}{R_1}$$

Solving for R_1 between the two equations for $I_{B(sat)}$ yields

$$R_1 = (V_1 - V_{BE(sat)})\, h_{FE(sat)} R_L / (V_{CC} - V_{CE(sat)})$$

or

$$R_1 = \frac{V_1}{V_{CC}} h_{FE(sat)} R_L$$

As shown earlier, the voltage representing logic '1' may have any value between V_{CC} and $V_{CC}/2$, or even lower. Thus the **minimum** value of R is needed when $V = V_{CC}/2$, that is when $R_1 = R_L h_{FE(sat)}/2$. When $V_1 = V_{CC}$, it is possible to use a value of $R_1 = R_L h_{FE(sat)}$. The former value is chosen since it allows the transistor to saturate when the lowest logic '1' signal is applied. In practice, it is also necessary to allow for variations not only in the supply voltage, but also in the tolerances of R_1 and R_L as well as changes in $h_{FE(sat)}$ between transistors. To allow for all these factors, the value chosen for R_1 may only be about 30 per cent of the maximum value, that is, 30 per cent of $R_L h_{FE(sat)}$. To illustrate the design procedure, we shall design a NOT gate which uses a supply voltage of 10 V, and in which the maximum collector current is to be 5 mA and the value of $h_{FE(sat)}$ is 20. Using the relationships deduced above

$$R_L = V_{CC}/I_{C(sat)} = 10\text{ V}/5\text{ mA} = 2\text{ k}\Omega$$

Since we have specified a *maximum* value of collector current of 5 mA, it is advisable to select a value of R_L which is the next preferred value above 2 kΩ, that is, $R_L = 2.2$ kΩ.

The preferred range of 10 per cent tolerance resistors are decimal multiples and sub-multiples of the following range: 10, 12, 15, 18, 22, 27, 33, 39, 43, 47, 56, 68, and 82.

If we can assume that the logic '1' voltage is always 10 V, then we may use a value for R_1 of

$$R_1 = h_{FE(sat)} R_L = 20 \times 2.2 = 44\text{ k}\Omega$$

If the output of the gate is loaded so that the logic '1' level falls to $V_{CC}/2$, then R_1 will need to have a value of 22 kΩ if the transistor is to be saturated. Furthermore, if we make allowances for all the factors mentioned above, we really need to select a value for R_1 of about

$$R_1 = 0.3 \times 44 = 13.2 \text{ k}\Omega$$

Using the next *lower* preferred value, we would select

$$R_1 = 12 \text{ k}\Omega$$

4.7 'Current-sourcing' gates and 'current-sinking' gates

The popular ranges of BJT logic gates incorporate diodes in their input circuits and, depending on the connection of the diodes, this affects the operation of the circuit. The two principal types of circuit involved are shown in figure 4.9.

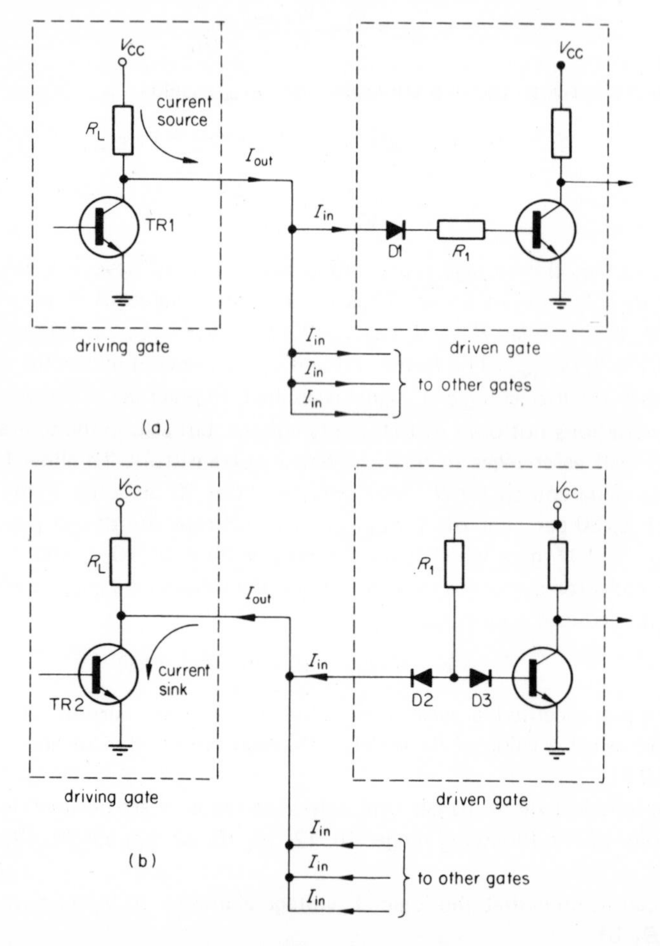

Fig. 4.9 Illustrating (a) a current 'sourcing' gate and (b) a current 'sinking' gate

In figure 4.9(a), when the output of the driving gate is logic '1', it acts as a **current source** which saturates the transistor in each of the driven gates. When the output of the driving gate is '0', then the anode of D1 is earthed and no current flows in the driven transistors. This type of circuit is typical of NOR gates.

Figure 4.9(b) is typical of NAND circuit elements, in which diode D2 is connected so as to cause the transistor in the driving gate to **sink** the current which flows in R_1 in the driven gate. Diode D3 is included in the circuit for voltage shifting purposes, and the reasons for its use is described more fully in chapter 5. Hence, when the output of the driving gate is logic '0', it acts as a **current sink** for the current which normally flows into the base of the transistor in the driven stage.

4.8 Fan-out

The fan-out capability of either current sinking or current sourcing gates is the maximum number of basic gate inputs it may simultaneously excite, without causing either of the logic levels to fall outside its specific range.

In the case of NOR-type gates (figure 4.9(a)), we are concerned with preventing the logic '1' level from falling below its specified minimum value. In NAND-type gates (figure 4.9(b)), the driving transistor must 'sink' the sum of the individual currents flowing in the inputs of the driven gates, without causing the driving transistor to come out of saturation. If the transistor were to come out of saturation, the logic '0' voltage would rise above its specified maximum value.

Let us consider a NOR network of the type in figure 4.9(a), which uses components with the values deduced in section 4.6, namely $R_1 = 12\ \text{k}\Omega$, $R_L = 2.2\ \text{k}\Omega$. If the fan-out of the basic gate is N then, when TR1 if OFF, the *minimum output voltage* at the collector of TR1 must be sufficient to saturate N other transistors. In order to evaluate N we need to specify the minimum voltage which represents logic '1'. Using $V_{CC} = 10$ V, let us assume that this value of voltage is 5 V. Assuming that we can neglect the p.d. in diode D1 when it is forward biased, and that $V_{BE(sat)}$ is negligible, then

$$I_{in} \approx 5/R_1 = 5/12\ \text{k}\Omega \approx 0.42\ \text{mA}$$

and

$$I_{out} = (10-5)/R_L = 5/2.2\ \text{k}\Omega \approx 2.3\ \text{mA}$$

Now

$$\text{fan-out} = N = \frac{I_{out}}{I_{in}} = \frac{2.3}{0.42} = 5.5$$

The operating fan-out is an integral value, so that we would specify a fan-out of 5 for a NOR-type gate of this kind.

In the NAND network of figure 4.9(b), the fan-out is once more given by $N = I_{out}/I_{in}$. However, we must bear in mind that the driving transistor TR2 must not only carry the current in R_L, but must also 'sink' a current of NI_{in}. It is therefore possible to use a higher value of R_L in the NAND circuit than is used in the NOR circuit.

4.9 Fan-in

The fan-in of a gate refers to the maximum number of separate input lines that may be used to control the state of the gate. The actual number of input lines included in the construction of the gate is frequently less than the theoretical maximum fan-in, since few applications call for more than about four input lines.

Where a large fan-in is required, it may be possible to use *fan-in expanders* (see chapter 5). In many cases the maximum fan-in is determined by the switching speed of the gate, since one limit on the switching speed is the capacitance of the input circuits.

4.10 Thermal considerations

Temperature changes affect the fan-in, the fan-out, the switching speed, and the logic levels of the gates. Gates used in industrial and commercial applications are designed to operate within their specified limits over a temperature range 0 to 75° C. Military systems and certain industrial systems which need a wider operating temperature range, use families of logic gates which have an operating temperature range from –55° to 125° C.

4.11 Noise immunity

The noise immunity of a logic gate is the degree to which it can withstand variations in logic levels at the input without causing the output state of the gate to change significantly.

The d.c. noise margins are defined in terms of the transfer characteristic of the gate, an example being shown in figure 4.10. In the characteristic shown, the HIGH output voltage is 3 V and the LOW output voltage is 0.2 V. Point A on the characteristic corresponds to the operation of the circuit when the input is energised by a LOW signal (0.2 V), and point B to the operating state when the input is HIGH (3 V). In the case considered, the transition from one logic level to the other occurs between input voltages of 0.9 V and 1.1 V. A positive-going 'noise' signal of (0.9 – 0.2) = 0.7 V can be superimposed on the minimum logic '0' input state without the circuit malfunctioning. Also, a negative-going 'noise' signal of (3 – 1.1) = 1.9 V can be superimposed on the maximum logic '1' input state without serious effect. However, if the driving gate is loaded so that the logic '1' signal applied to the input is only 1.5 V, then the noise margin for the HIGH input state is only 0.4 V.

Similarly, the LOW input noise margin is affected by the maximum allowable '0' voltage level. If, for example, this is 0.4 V, then the LOW input noise margin is only (0.9 – 0.4) = 0.5V.

4.12 Time delays in a BJT switch

The speed with which a signal can be propagated through an electronic gate depends both on the number of time delays involved and also on their magnitudes. Typical switching waveforms associated with a NOT gate are shown in figure 4.11.

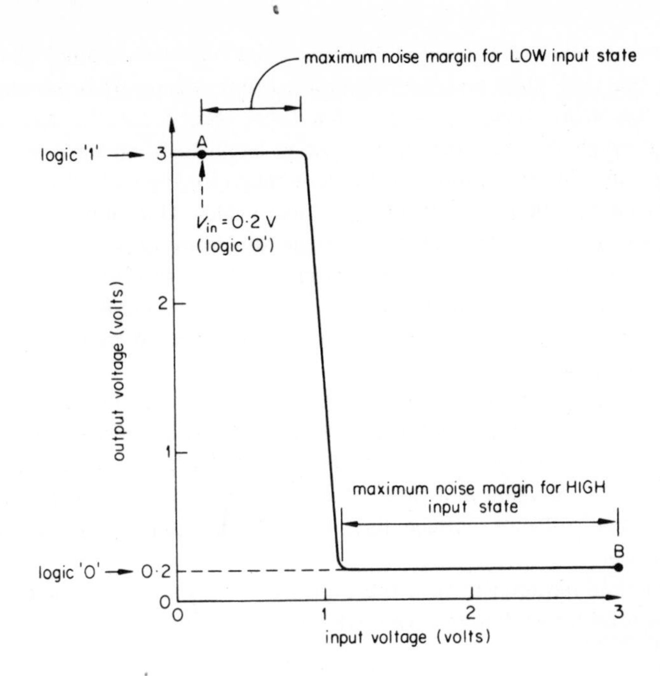

Fig. 4.10 A typical transfer characteristic for an INVERTING gate

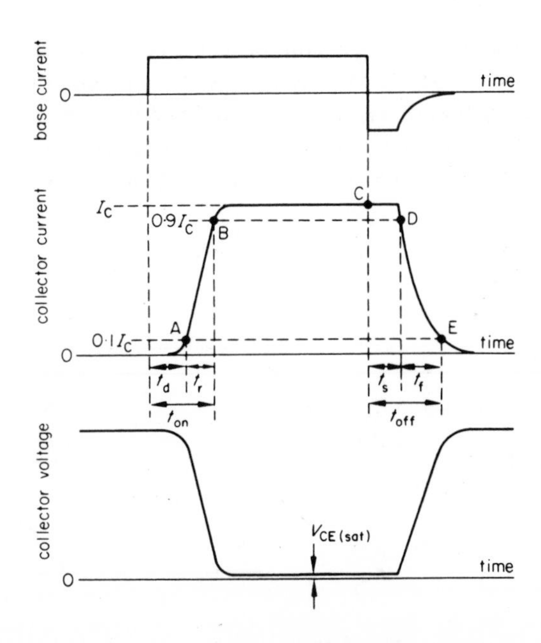

Fig. 4.11 Switching waveforms of an INVERTING gate

For a short period of time after the base drive has been applied, the transistor remains in the OFF state and the collector current is zero. This period of time is known as the **delay time,** t_d, and is the time needed for the base current to propagate through the base region in order to begin to bring the transistor into a conducting state. The **rise time,** t_r , is the time taken for the collector current to rise from 10 per cent to 90 per cent of its maximum value. During this interval of time and with a resistive load, the collector voltage falls from 90 per cent to 10 per cent of its maximum value. The rise time is determined by several factors including the high frequency response of the transistor, its current gain, and the magnitude of the base drive. Transistors used in switching applications should have as high a cut-off frequency as possible.

When the base voltage is reduced to zero, the collector current remains constant for a period of time known as the **storage time,** t_s, which is the time required to remove excess base charge. The collector current falls from 90 per cent to 10 per cent of its maximum value in a time interval known as the **fall time,** t_f, and during this interval the collector voltage rises from 10 per cent to 90 per cent of its maximum value.

The total time required to turn the transistor ON is known as the **turn-on time,** t_{on} , and the total time taken to turn it OFF is the **turn-off time,** t_{off}, where

$$t_{on} = t_d + t_r$$

and

$$t_{off} = t_s + t_f$$

Typical values for a silicon switching transistor are t_{on} = 12 ns, t_{off} = 15 ns, with t_s = 10 ns.

An important factor in specifying the switching performance of gates is the time taken for a signal to propagate through the gate from its input to its output, and is known as the **propagation delay,** t_{pd} , and is defined in terms of the waveforms in figure 4.12(b). The waveforms relate to the inverting gate in figure 4.12(a). The propagation delay is specified at the 50 per cent voltage levels of input and output signals, where

$$t_{pd} = \frac{t_1 + t_2}{2}$$

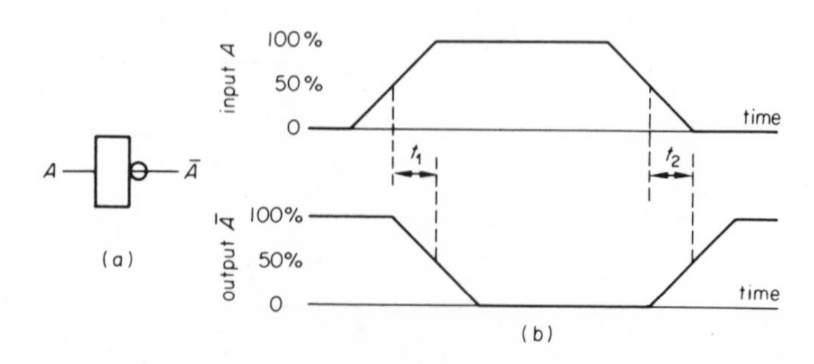

Fig. 4.12 Propagation delay

The propagation delay depends on many factors including the circuit design, the supply voltage, the power consumption, etc., and may have a value from about 1 ns to a fraction of a microsecond. Values for typical logic families are given in chapter 5.

4.13 Methods of improving the switching speed

Two methods used to reduce the switching time of gates are illustrated in figure 4.13. The first of these methods is to shunt the input resistor *R* with a **speed-up capacitor** *C*. This allows the total input voltage transition to be momentarily applied to the base of the transistor. Thus, when signal *A* changes from zero to a positive voltage, the capacitor charging current causes a momentary rush of current to be applied to the transistor base, so reducing the delay time. When signal *A* is reduced to zero, the negative-going voltage transition is transmitted to the transistor base. This has the effect of reducing the storage time.

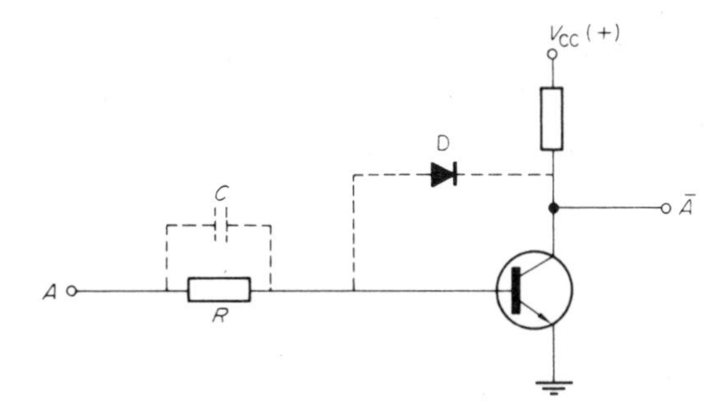

Fig. 4.13 Methods of improving the switching speed of an INVERTING gate

Unfortunately, the speed-up capacitor allows noise signals to be transmitted to the transistor base, so that it reduces the noise margin to some extent. In many cases, in order to obtain an improved switching performance, it is advisable to use a transistor with a higher cut-off frequency rather than to use a speed-up capacitor.

As we have seen, a factor which limits the switching speed is the storage time of the transistor. The storage time can be reduced by preventing the transistor from being driven too hard into saturation, that is, by limiting the base current to a value which is just sufficient to saturate the transistor. One method of achieving this solution is by the use of the **clamping diode** D. The onset of saturation occurs when the collector voltage falls below the base voltage; when this occurs diode D becomes forward biased, and the excess base current is diverted via diode D and the transistor to earth. Since the diode only carries the excess base current, the storage time of the diode is much shorter than that of an overdriven transistor.

Ideally, the clamping diode should have zero storage time which is obtained in IC logic gates which use Schottky diodes (see section 5.5).

4.14 Active collector loads

An **active** circuit element is one which provides voltage gain or current gain, examples of which are bipolar transistors and field-effect transistors (see section 4.16). A **passive** circuit element is one which does not exhibit the property of gain, examples of which are resistors, capacitors, and inductors.

In the NOT circuit described earlier, a passive resistor (often referred to as a **pull-up resistor**) was used as the collector load of the transistor. In order to change the value of the current flowing through the resistor, it is necessary to either charge or discharge the stray capacitance associated with that resistor. If the resistor is replaced by an **active load** then, as a result of the gain of the active device, the rate at which the output voltage can change is accelerated. One form of active load is illustrated in figure 4.14. Here the active load comprises transistor TR2 (the **pull-up transistor**), diode D, and resistor R. The circuit operates as follows. When $X = 1$, transistor TR1 is saturated and the output voltage is LOW, that is, logic '0'. The presence of diode D in the circuit causes the emitter voltage of TR2 to be higher than its base voltage, so that TR2 is cut-off. When $X = 0$, transistor TR1 is OFF, so that the output voltage depends on the operation of the active load In this condition, current flows into the base of TR2 via resistor R, causing TR2 to

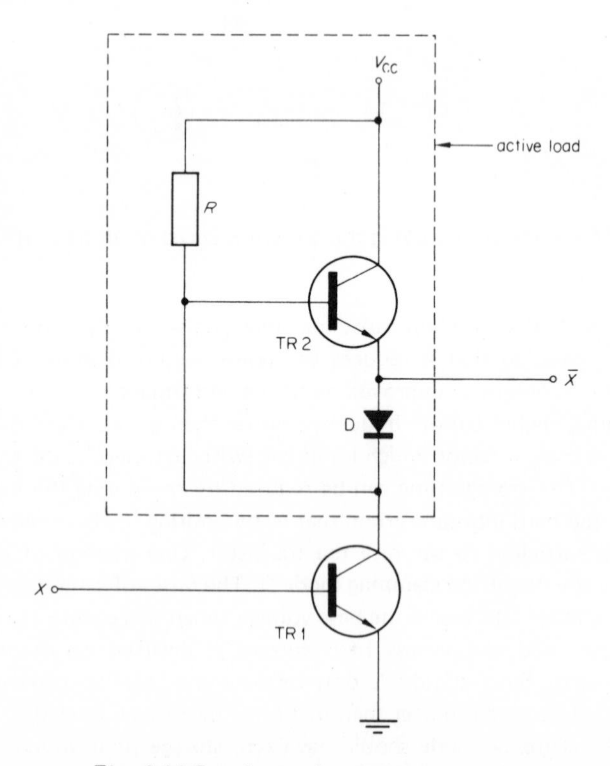

Fig. 4.14 One form of active collector load

saturate. Thus the voltage drop between the supply line and the output terminal is the $V_{CE(sat)}$ of TR2, that is, the output voltage is very nearly equal to V_{CC}. Hence, we see that an input of logic '0' gives a logic '1' output whose value is very nearly equal to V_{CC}.

Diode D fulfills the important function of preventing both TR1 and TR2 from being switched ON simultaneously. If it were possible for the two transistors to be turned ON together, even for a few nanoseconds, the resulting current spike would generate a great deal of electronic noise and would also cause excessive power dissipation in the circuit.

4.15 The wired-OR function

It is convenient in many cases to generate logical functions by using what is known as the wired-OR connection of logic gates, in which the output terminals of the gates are connected together. The general arrangement is shown in figure 4.15(a), with the symbolic representation in (b). Although NOR-type invertors are shown in (b), this connection can, with the exception given below, also be used with NAND gates.

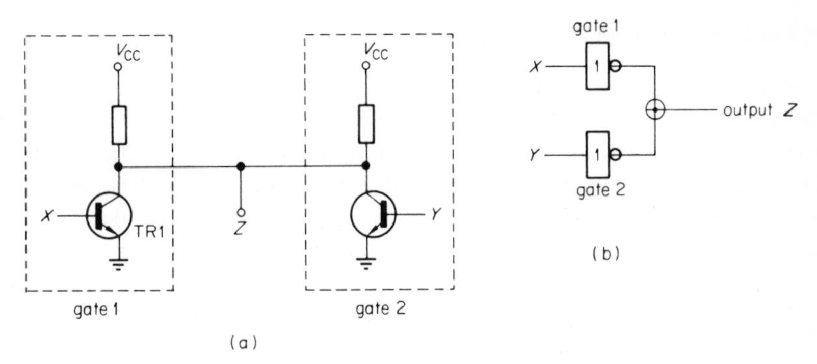

Fig. 4.15 The wired-OR connection

In figure 4.15(a), when the signal $X = 1$ then TR1 is saturated so that $Z = 0$ irrespective of the logical value of Y. Similarly, when $Y = 1$ then $Z = 0$ irrespective of the value of X. The truth table for figure 4.15(a) is given in table 4.1.

Table 4.1

X	Y	Z
0	0	1
0	1	0
1	0	0
1	1	0

Thus $Z = 1$ when $X = 0$ and $Y = 0$, that is

$$Z = \bar{X} \cdot \bar{Y}$$

Hence the logical output from the circuit is given by the relationship

Z = NOT (the function at the base of TR1) AND NOT (the function at the base of TR2)

In the general case, the functions generated at the bases of the transistors are fairly complex, so that the overall function is usually more complex than that given above. For example, if $X = A \cdot B$ and $Y = C \cdot D$, then

$$Z = \overline{A \cdot B} \cdot \overline{C \cdot D} = (\bar{A} + \bar{B}) \cdot (\bar{C} + \bar{D})$$

An important point to note with this connection is that both gates use passive pull-up resistors. If active pull-up transistors are used then, when $X = 1$ and $Y = 0$ or when $X = 0$ and $Y = 1$, a continuous short-circuit is applied to the power supply via the wired-OR link. **It is general practice, therefore, to use only gates with resistive loads in wired-OR networks.**

4.16 Insulated-gate field-effect transistors

Insulated-gate field-effect transistors (IGFET's) depend for their operation on the influence of an electric field on the conductivity of a very thin region of semiconductor material known as a conducting **channel.** A cross-section through an *insulated gate p-channel FET* is shown in figure 4.16(a). The **drain electrode** is maintained at a negative potential with respect to the **source electrode** so that, with zero gate voltage, the *p–n* junction between the drain and the substrate is reverse biased. The initial drain current is therefore zero, and this lack of conductivity is represented in the circuit symbol, figure 4.16(b), by breaks in the link between the source (S) and the drain (D). The construction of the device also leads to it being described as a **MOSFET** (**m**etal **o**xide **s**emiconductor **FET**), since the gate-to-channel construction has a metal oxide semiconductor geometry.

The application of a negative potential to the gate attracts any free *p*–type charge carriers (that is, holes) in the substrate to the underside of the oxide immediately below the gate. At a voltage known as the **threshold voltage,** V_T, a sufficient number of positive charge carriers have collected at the oxide-to-semiconductor interface to form a conducting channel which links the source and drain. The value of V_T lies, typically, between about −2 V and −5 V. This is shown in the mutual characteristic of the MOSFET in figure 4.16(c). Increasing the negative gate voltage causes the conductivity of the channel to increase, so that the drain current increases in the manner shown in figure 4.16(d).

Several types of FET are manufactured, the most popular type for use in logic circuits being the *p*-channel MOSFET described above. The principal features which make *p*-channel MOS (*p*-MOS) devices more attractive in some instances than bipolar transistors in logic applications are listed below.

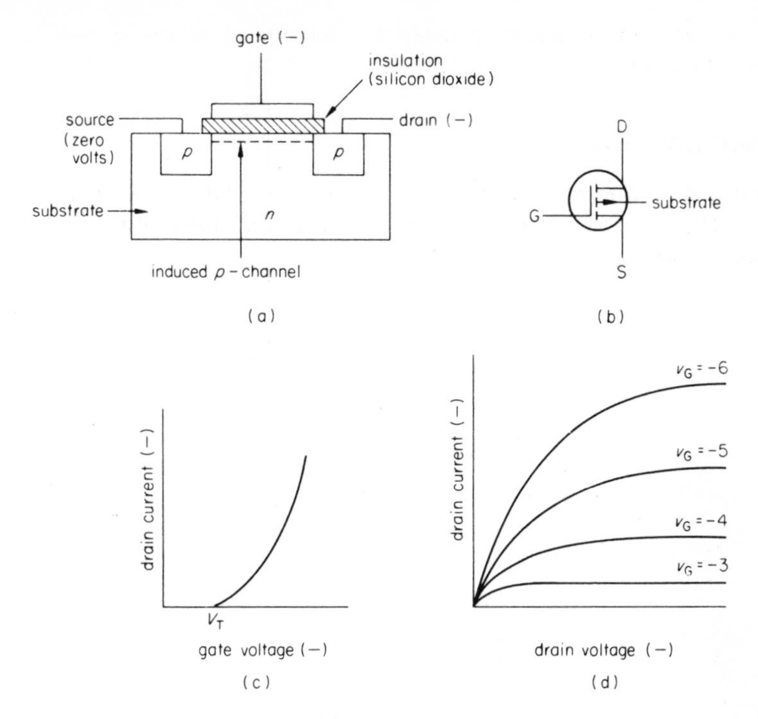

Fig. 4.16 (a) Sectional view of a *p*-channel insulated-gate FET, (b) its circuit symbol, (c) typical gate characteristics and (d) typical common-source output characteristics.

1. MOS devices are physically smaller than bipolar transistors, giving a reduction in the cost per gate.
2. Their construction is compatible with monolithic integrated circuit production techniques.
3. Their input resistance is very high, typically $10^{12}\Omega$.
4. MOS devices can be used to replace resistors, with a saving both in size and cost. In this mode, MOS devices are described as **pinch-effect resistors.**
5. The output from the driving gate can be directly connected to the inputs of the driven gates without need for bias networks.

A disadvantage of MOS devices when compared with bipolar devices is their lower switching speed. This is due largely to the input capacitance associated with the oxide layer which separates the gate from the conducting channel.

MOSFET's do not exhibit the charge storage effects associated with BJT's, but they do display time lags due to interelectrode capacitance. The output waveforms are generally similar to those shown in figures 4.11 and 4.12, with the exception that t_d in figure 4.11 is replaced by the **turn-on delay,** $t_{d(on)}$, of the FET, and t_s in figure 4.11 is replaced by the **turn-off delay,** $t_{d(off)}$, of the FET. The rise time, the fall time, and the propagation delay are determined in the manner shown in figures 4.11 and 4.12.

4.17 MOS NOT gates

A *p*–MOS NOT gate is shown in figure 4.17, in which TR2 is used as a pinch-effect resistor. The principle of operation is generally similar to that of the BJT NOT gate described in section 4.6.

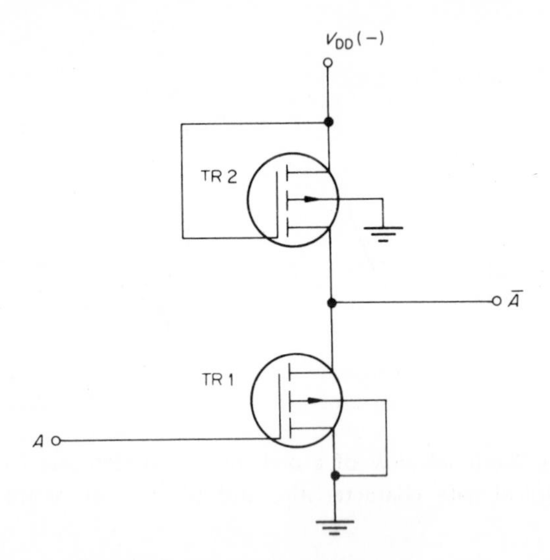

Fig. 4.17 A *p*-channel MOS NOT gate

P–MOS gates require a negative supply voltage, so that with $V_{DD} = -20$ V, the HIGH output voltage lies between about −14 V and −11 V. The LOW output voltage is between about −2 V and −3 V. For convenience the logic '1' level is represented by the more negative voltage, that is, −11 V to −14 V. That is, ***p*–MOS gates operate in negative logic.**

The basis of another family of MOS gates is shown in figure 4.18. This circuit employs **complementary MOS transistors,** TR1 being an *n*–channel device and TR2 being a *p*–channel device. This type of gate is known as a COSMOS gate or CMOS gate. The principal advantages claimed for CMOS gates over *p*–MOS are an improvement in switching speed of two to three times, and a significant reduction in power consumption. Offset against this is the fact that CMOS gates take up a greater surface area on the semiconductor chip than do *p*–MOS, and that more processes are required in their production. CMOS gates work with a positive supply voltage and operate in *positive logic.* A logic '0' signal applied to the input causes TR1 to turn OFF and TR2 to turn ON. A logic '1' input signal results in TR1 turning ON and TR2 turning OFF. Due to the mode of operation of the circuit, the

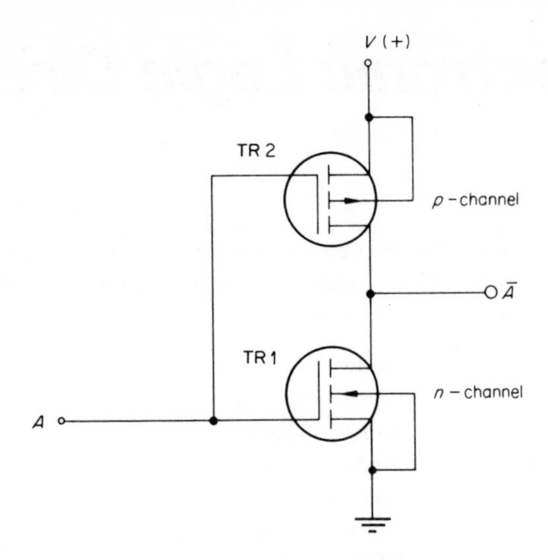

Fig. 4.18 A complementary MOS NOT gate

HIGH output voltage is almost equal to V_{DD} (which lies in the range 3–15 V, and the LOW output voltage is practically zero (typically 10^{-2}V). Also, as TR1 is ON when TR2 is OFF and vice versa, the quiescent power consumption per gate is very small, being typically 10^{-8}W.

Due to the very high input impedance of these gates, the input current per gate is extremely low (typically 10 pA). A fan-out in excess of 1000 can be achieved at low operating frequencies (that is, below about 10 kHz).

5 Electronic Logic Circuits

In this chapter we shall deal with all the popular digital logic families and, finally, shall outline the important constructional techniques associated with their production. The principal logic families are:

DRL	diode–resistor logic
RTL	resistor–transistor logic
DCTL	direct-coupled transistor logic
DTL	diode–transistor logic
TTL	transistor–transistor logic
ECL	emitter-coupled logic
p-MOS and CMOS	MOS logic families

5.1 Diode–resistor logic (DRL)

DRL is a basic logic family and, as a group of circuits, plays little part in logic systems as we know them today. However, the principles involved in their operation are of great importance to the **integrated circuits (IC)** which follow.

OR gate

A basic 2-input OR gate circuit is shown in figure 5.1, and consists of two diodes and a resistor. With the input switches in the positions shown, that is, $X = Y = 0$, the net e.m.f. acting around the circuit is zero, so that $Z = 0$.

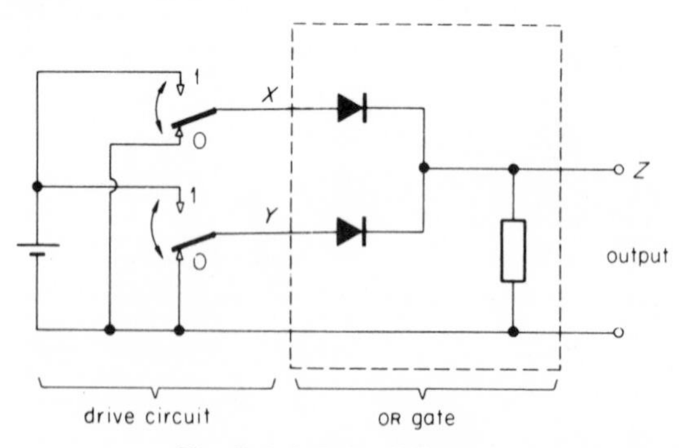

Fig. 5.1 A DRL OR gate

When either $X = 1$ or $Y = 1$, or when both inputs are at the logic '1' level, one or both of the diodes are forward biased so that $Z = 1$. This operation satisfies the OR function truth table.

AND gate

In figure 5.2, when either input is switched to the logic '0' position the diode in that line is forward biased, thereby connecting the output line to zero potential.

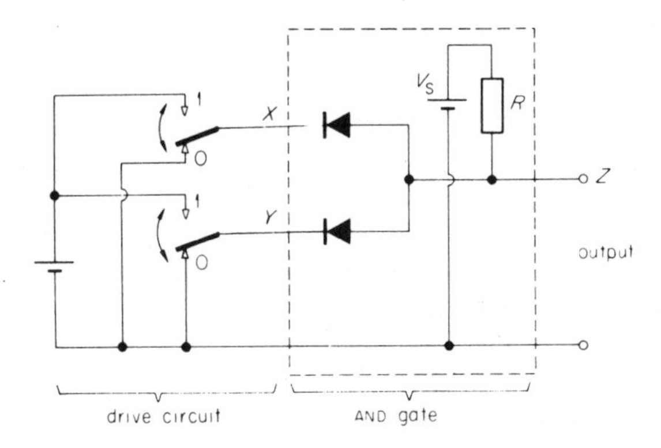

Fig. 5.2 A DRL AND gate

That is, when either input is logic '0', then the output is also logic '0'. Only when both inputs are at the logic '1' level are the diodes cut off. When the circuit is in this state, the current in R falls to a low value and the output voltage rises to the logic '1' level.

5.2 Resistor– transistor logic (RTL)

RTL represented an early stage in the development of digital logic engineering and one circuit, the NOT gate, was discussed in chapter 4. In thes family, bipolar NOT gates are used in conjunction with versions of DRL OR and AND gates to give overall NOR and NAND functions, respectively. The circuits were constructed from individual components and, consequently, were expensive. Also, their propagation speed was limited by the types of transistor available at that time. These circuits have been superseded by the integrated circuit types described in the following sections.

5.3 Direct-coupled transistor logic (DCTL)

The DCTL gates shown in figure 5.3 were the first types to be manufactured in monolithic integrated circuit form (see also section 5.8), and represented a significant step forward in digital electronics.

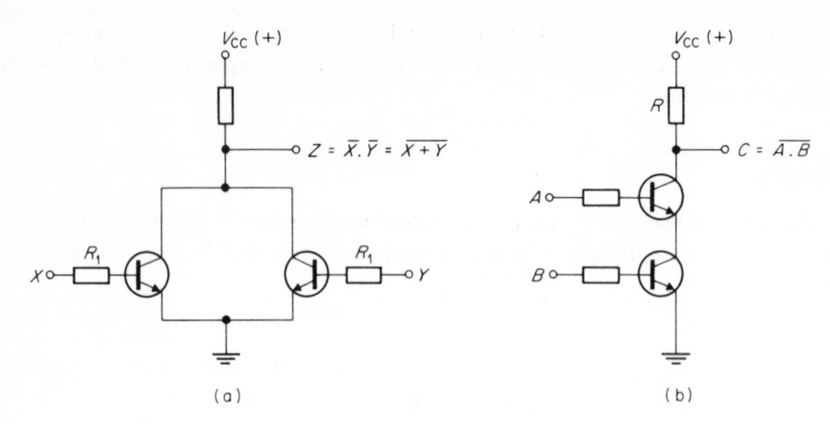

Fig. 5.3 (a) DCTL NOR gate and (b) DCTL NAND gate

The NOR gate, figure 5.3(a) uses the wired-OR connection of two transistors and, as was shown in section 4.15, the output is

$$Z = \overline{X} \,.\, \overline{Y}$$

It is shown in chapter 6 (see De Morgan's theorem) that this may be rewritten in the form

$$Z = \overline{X + Y}$$

That is, the output from figure 5.3(a) is the NOR function of the inputs.

Owing to dissimilarities between transistor input characteristics, it is necessary to include a resistor R_1 in series with the base of each of the transistors in order to limit the maximum base current which may be drawn by each transistor. Otherwise one input may **hog** all the driving current.

The NAND circuit, figure 5.3(b), employs two transistors in series. Whenever either of the transistors is OFF, no current flows through R and the output voltage is HIGH, that is, logic '1'. That is, when either input signal is logic '0', the output is logic '1'. Only when both input signals are logic '1' is the output line connected to ground, causing the output signal to be logic '0'.

Although DCTL gates have largely fallen into disuse, the basic circuits in figure 5.3 are used in MOS logic gates (see section 5.7).

Both gates in figure 5.3 are *current sourcing* gates, since the gate acts as a current source when the output voltage is HIGH.

Typical performance parameters of this family of gates are

supply voltage	3.6 V
logic '1'	1.6 V (minimum)
logic '0'	0.2 V
propagation delay	25 ns
fan-out	5

5.4 Diode–transistor logic (DTL)

A basic form of DTL NAND gate is illustrated in figure 5.4, and consists of a DRL AND section followed by an invertor. The function generated at point Y is the AND function of inputs *A* and *B*, and is inverted by the transistor output stage.

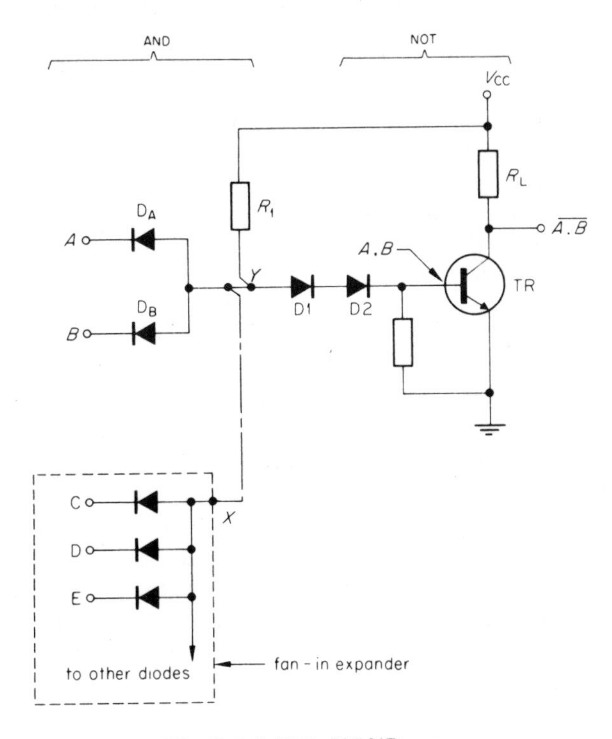

Fig. 5.4 A DTL NAND gate

When either input *A* or input *B* is logic '0', the circuit designer must ensure that the current flowing in R_1 chooses to flow through the appropriate input diode (D_A or D_B), rather than through D1 and D2. If this were not the case, the current flowing into the base of the transistor would not fall to zero when either input is zero. A preferential bias is given to the flow of current for a logic '0' input by including the two **voltage offset diodes** D1 and D2 in series with the base. The forward biased potential difference across these diodes, together with $V_{BE(sat)}$ of TR, exceeds the sum of the maximum logic '0' level and the forward potential difference in either D_A or D_B.

DTL gates of the type described are *current sinking gates* within the terms of the definition in section 4.7. The driving transistor is required to 'sink' the current flowing in R_1 in each of the driven gates.

The fan-in of the circuit in figure 5.4 can be increased by connecting the diode

fan-in expander shown in the inset to the figure, points X and Y being linked so that the logic function generated at point Y is $A \cdot B \cdot C \cdot D \cdot E \cdot \ldots$

Since the circuit in figure 5.4 employs a passive pull-up resistor R_L, it can be used in conjunction with other identical gates in the wired-OR configuration. Certain types of DTL NAND gates employ active pull-up loads similar to the one shown in figure 4.14. The latter gates should not be used in wired-OR networks for the reasons given in section 4.14.

For the circuit shown in figure 5.4, the following values are typical.

supply voltage	6 V
typical noise immunity	1.2 V
propagation delay	30 ns
power per gate	11 mW
logic '1'	4 V (minimum)
logic '0'	0.4 V (maximum)
fan-out	8
fan-in	14

A modified version of the NAND gate which is better suited to integrated circuit production techniques is shown in figure 5.5. This circuit provides improved fan capability together with a smaller propagation delay when compared with figure 5.4. The improved performance is brought about by replacing D1 in figure 5.4 by transistor TR1 in figure 5.5. In figure 5.5, the driving gate has merely to 'sink' the base current of TR1, rather than all the current flowing in R_1.

In an industrial environment it is desirable to have the highest possible value of noise immunity. One version of DTL with a high noise immunity includes the modification shown in the inset in figure 5.5. Diode D2 in the figure is removed, and the circuit shown in the inset is connected to the points marked L, M, N. The

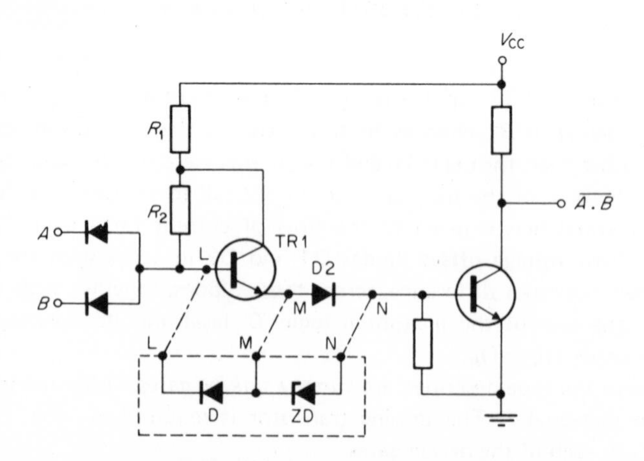

Fig. 5.5 An integrated circuit DTL NAND gate

Zener diode ZD provides the improved noise immunity, which has a value of approximately $(V_Z + 0.7)$ V, where V_Z is the breakdown voltage of ZD. Zener diodes have a relatively large parasitic capacitance, and diode D is included to provide a discharge path for the charge held by this capacitance when the input signal changes to the '0' state.

To avoid the possibility of pick-up voltages on unconnected inputs causing system malfunction, it is usual either to connect spare inputs to lines already in use or to connect them to the supply line. This practice should also be adopted with TTL gates (see section 5.5).

5.5 Transistor–transistor logic (TTL)

The development of TTL represented a break from conventional designs, and was made possible by developments in monolithic IC production techniques. The basic form of TTL gate, a NAND gate, is shown in figure 5.6 and uses a **multi-emitter transistor** TR1 in the input circuit. Transistor TR2 is used in a phase-splitting amplifier, and provides complementary logical signals at its emitter and collector.

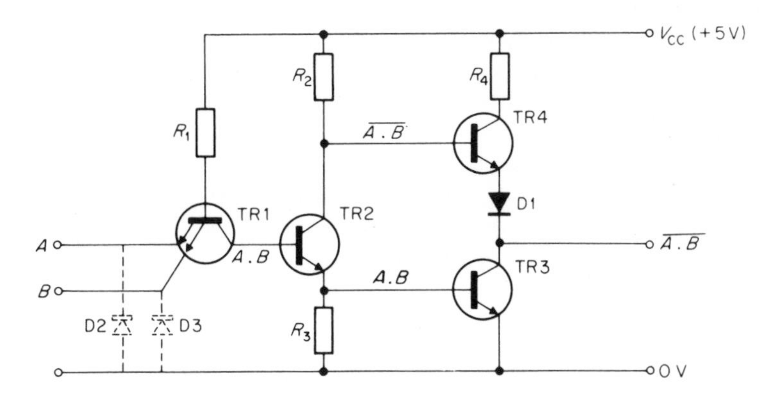

Fig. 5.6 A TTL NAND gate

Transistor TR4 acts as an active load for transistor TR3, the two being energised by the complementary outputs from TR2. The circuit operation is described below.

When any input line is at the logic '0' level, the current flowing in R_1 is diverted to that line. The flow of base-emitter current causes TR1 to saturate, thereby causing TR1 collector voltage to fall to the logic '0' level; that is, the collector signal of TR1 is the AND function of the inputs. This LOW voltage is applied to the base of TR2 and causes it to be cut-off, so that its emitter voltage is LOW and its collector voltage is HIGH. In turn, this results in TR3 being turned OFF and TR4 being ON, connecting the output line to the supply voltage. That is, a logic '0' on any input causes the output to be logic '1'.

When all inputs are at logic '1', the current flowing in R_1 is diverted through the collector region of TR1 into the base of TR2, resulting in TR2 saturating. This immediately causes the emitter voltage of TR2 to rise, and its collector voltage to fall. The HIGH voltage at the emitter of TR2 drives TR3 into saturation, and connects the output terminal to the zero volts line. Also, the reduced collector voltage of TR2 causes TR4 to be cut-off, so that no current flows through TR4.

The reason for the use of diode D becomes apparent when we study the quiescent voltages existing in the circuit in the above operating state. Since, in a silicon transistor, $V_{BE(sat)}$ is about 0.7 V and $V_{CE(sat)}$ is about 0.2 V and, in a silicon diode, the forward biased potential difference is about 0.7 V, the following voltages exist with all inputs HIGH.

at the base of TR3 (emitter of TR2)	0.7 V
at the base of TR4 (collector of TR2)	0.9 V
at the collector of TR3	0.2 V

Thus, the potential between the base of TR4 and the collector of TR3 is only about 0.7 V, which is insufficient to cause TR4 and D1 to conduct. Had D1 not been included, this voltage would be sufficient to maintain TR4 in a conducting state. The output circuit shown in figure 5.6 is sometimes known as a **totem pole** circuit because of its shape.

When the output voltage is caused to change from one level to the other, both TR3 and TR4 may conduct simultaneously for a very short interval of time, resulting in a surge of current being drawn from the supply. Resistor R_4 is included to limit the value of this surge.

TTL gates dominate the IC logic market, the family having three basic branches which are known as **low-power, standard,** and **high-speed.** The standard version is the most popular, the essential difference between the three groups being the values chosen for R_1, R_2, R_3, and R_4. Typical values are given in table 5.1.

Table 5.1

	R_1 (kΩ)	R_2 (kΩ)	R_3 (kΩ)	R_4 (Ω)	power (mW)	propagation time (ns)
low-power	40	20	12	500	1	30
standard	4	1.6	1	130	10	12
high speed	2.8	0.76	0.47	58	20	3 to 6

With a 5 V supply, the normal logic '1' level is about 3.3 V (minimum value about 2.4 V). The maximum logic '0' level is 0.4 V, and the fan-out is about 10.

The circuit propagation delays are primarily determined by the rate at which the inherent capacitances of the components can be charged and discharged. To speed up the charging and discharging process the transistors must provide a higher current, which is obtained by reducing the values of the circuit resistors.

The switching speed is also increased by the use of a clamping technique which prevents the output transistor from being driven too far into saturation. A method mentioned earlier is the use of a diode clamp between the base and the collector (see section 4.13). A Schottky diode (see section 4.4) is formed between a metal connection and the *n*-type silicon of the collector region of the transistor, the collector region acting as the cathode of the diode. The operation of the Schottky diode is temperature dependent, and as the temperature rises so the Schottky clamp becomes less effective and allows the transistor to be driven further into saturation. Moreover, the noise immunity of Schottky-clamped gates is less than other forms of TTL gates. Even so, the improved performance of Schottky-clamped gates outweighs the disadvantages associated with reduced noise immunity and thermal effects.

The very fast rise and fall-times associated with TTL gates (about 1.5 ns/V) sometimes bring in their wake oscillations of voltage on the output line. Under certain circumstances, a voltage undershoot of greater than −2 V may be applied to the input of a driven gate. One solution sometimes adopted to overcome this problem is the use of the voltage **catching diodes** D2 and D3 shown in figure 5.6. Reflections can occur along any line when the rise or fall-time of the propagated wave is comparable with the time taken for a signal to travel the length of the line. An electrical signal travels at the rate of about 0.3 m/ns hence, since the rise-time of a TTL gate is about 5 ns, a line is regarded as being electrically long in a TTL system when its length is about 1.5 m.

In TTL networks, as with DTL networks, in order to avoid the possibility of spurious switching from the pick-up voltages on unused input lines, the spare input lines should either be connected to the +5 V line or connected in parallel with used inputs.

The basic TTL gate in figure 5.6 cannot be used in wired-OR networks because of the active load TR4. Special types of **open-collector** TTL gates, in which R_4, TR4, and D1 are omitted, are manufactured and can be used in conjunction with an external resistor in the wired-OR configuration.

A typical transfer characteristic for a TTL gate is shown in figure 5.7. Using a 5 V supply, the logic '1' output is constant at 3.3 V until the input voltage reaches 0.7 V (point A). At this point TR2 in figure 5.6 begins to conduct, but TR3 has not yet begun to carry current. Between points A and B on the characteristic, TR2 operates as part of a linear amplifier having a gain of $-R_2/R_3$ (that is, about −1.6 for the standard gate). When the input voltage reaches 1.4 V, at point B, TR3 begins to conduct, and the output voltage rapidly falls to $V_{CE(sat)}$ which is about 0.2 V.

Yet another version of TTL is the so-called **tri-state TTL gate,** which has a control line in addition to the normal inputs which enables the output impedance of the gate to be switched to a high value (typically 125 kΩ) when its output voltage is between about 0.4 and 2.4 V. This is an addition to the low output impedance states when the output is either '1' or '0'. This facility increases the versatility of the gate.

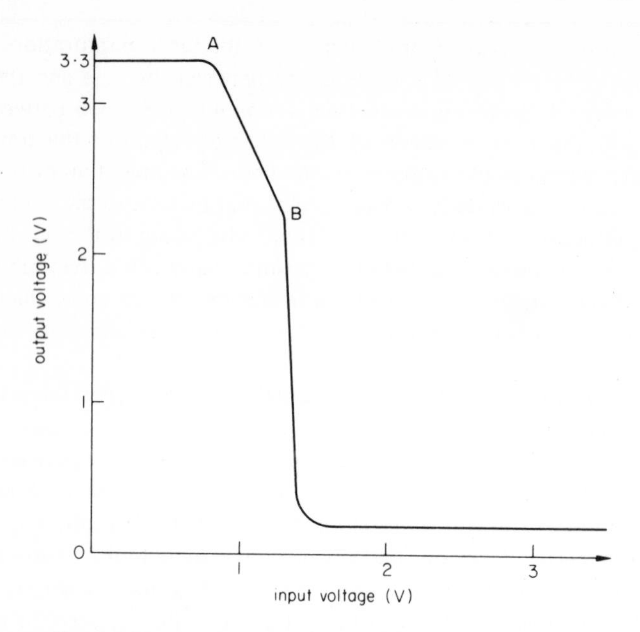

Fig. 5.7 Typical transfer characteristics of a TTL gate

5.6 Emitter-coupled logic (ECL)

The time delays associated with charge storage in the saturated logic gates described earlier can be eliminated if the transistors are not allowed to saturate. The ECL family of gates, with a typical OR/NOR gate being shown in figure 5.8, were developed with this purpose in mind. The propagation delay of this family of gates is typically less than 2.5 ns, and the fan-out is 30. Its drawbacks include a high power dissipation per gate (about 60 mW) and high sensitivity to temperature changes.

The basis of the circuit is a non-saturated emitter-coupled amplifier which consists of two sections, the right-hand section containing TR3 and the left-hand section containing TR1 and TR2. The value of R_E is large compared with both R_1 and R_2, so that the current flowing through the emitter-coupled amplifier is largely fixed by the value of R_E. The base of TR3 is supplied by a voltage reference source V_R of −1.15 V. The circuit operates with positive logic, logic '1' corresponding to an output of −0.75 V, and logic '0' to −1.55 V, giving a voltage swing of 0.8 V between the two levels. Thus, when $A = B = 0$, the current in R_1 is at its minimum value and in R_2 is at its maximum value. When either or both of the inputs are logic '1', the current in R_1 is increased and that in R_2 is reduced. Since TR1 and TR2 are connected in a wired-OR configuration, the output at the collector of TR1 is $\overline{A} \,.\, \overline{B}$ which, by De Morgan's theorem, is $\overline{A + B}$, that is, the NOR function of the inputs. As a result of the action described above, we see that the logical function at the collector of TR3 is $A + B$.

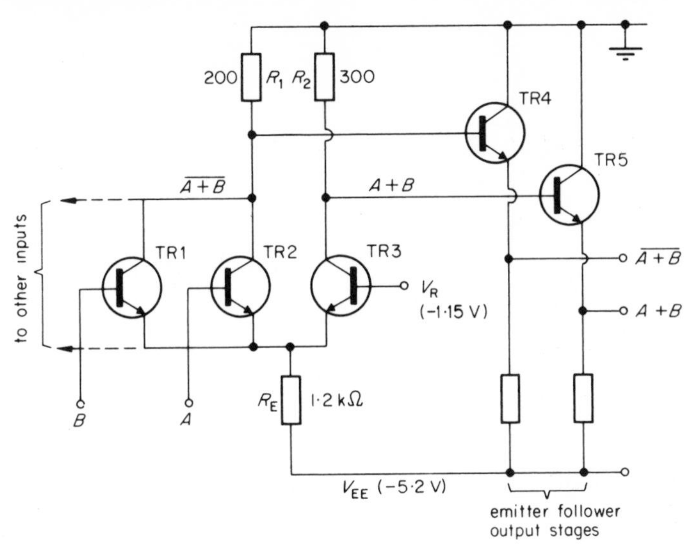

Fig. 5.8 An ECL OR/NOR gate

The emitter follower stages containing TR4 and TR5 fulfill two functions. Firstly, they provide a low impedance output which results in a large fan-out. Secondly, the V_{BE} drop of TR4 and TR5 restores the output voltages to the correct levels for driving other stages.

The following values are typical of an ECL OR/NOR gate.

supply voltage (V_{EE})	– 5.2 V
logic '1'	– 0.75 V
logic '0'	– 1.55 V
fan-out	25
propagation delay	2 ns

Other names used to describe ECL are *current-mode logic* (CML), *emitter–emitter coupled logic* (E^2CL), and *emitter-coupled transistor logic* (ECTL).

5.7 MOS logic gates

A *p*–MOS 2-input NOR gate using two MOSFET's in the wired-OR connection is shown in figure 5.9(a), and a 2-input NAND gate using series connected FET's is illustrated in figure 5.9(b). These circuits operate in negative logic, otherwise their operation is generally similar to the DCTL gates described in section 5.3. With a drain supply voltage V_{DD} of –20 V, the following values are typical of *p*–MOS gates.

propagation delay	100 ns
power per gate	7 mW for logic '0' output zero for logic '1' output
logic '1'	– 11 V (minimum)
logic '0'	– 3 V (maximum)

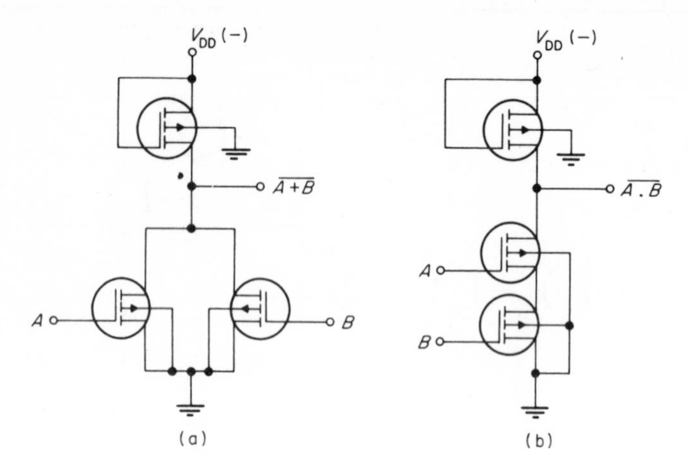

Fig. 5.9 (a) A NOR gate and (b) a NAND gate using *p*-channel MOS devices

The circuit in figure 5.10(a) is that of a 2-input positive logic CMOS NOR gate, and that in figure 5.10(b) is for a positive logic 2-input CMOS NAND gate. As outlined in chapter 4, the use of an *n*–channel device reduces the propagation delay and allows the maximum operating frequency of CMOS devices to be doubled when compared with that of *p*–MOS devices.

CMOS devices can operate over a very wide range of supply voltages, 3–15 V being typical. This allows CMOS gates to be operated from the same power source as TTL and DTL systems. The noise immunity of CMOS is about $0.45 V_{DD}$. The

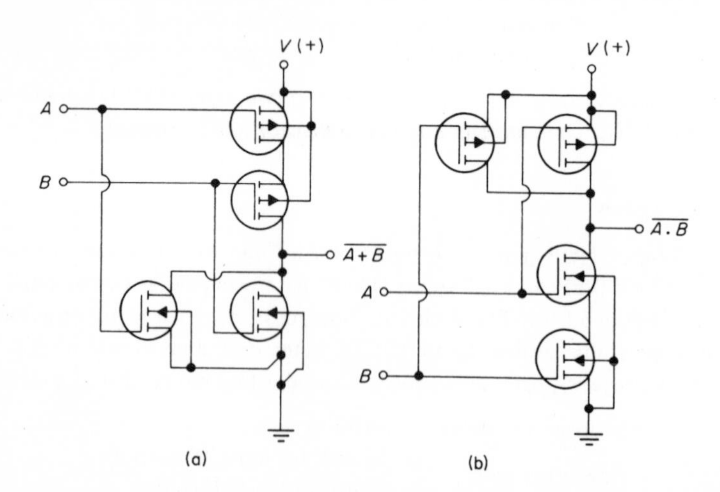

Fig. 5.10 (a) A NOR gate and (b) a NAND gate using complementary MOS devices

following figures apply to one family of CMOS gates when operating on a supply of 5 V.

noise immunity	2.25 V
propagation delay	40 ns
power per gate	5 nW
logic '1'	4.99 V (minimum)
logic '0'	0.01 V (maximum)

The very low quiescent power consumption arises from the fact that when the *p*–channel devices are ON then the *n*–channel devices are OFF, and vice versa. Also the fan-out is large since the current consumption per driven gate is only about 10 pA.

5.8 Monolithic integrated circuit construction

Integrated circuits used in electronic logic systems are usually constructed from silicon in what is known as **monolithic planar** form. That is, they are manufactured in a single slice of silicon in a 'flat' or plane form, and do not normally require the use of external components. For example, the integrated circuit shown in cross-section in figure 5.11(b) is a monolithic form of the bipolar circuit in figure 5.11(a).

In the following, a simplified version of the construction of figure 5.11 is given. First, the silicon is refined into a cylindrical ingot, which is then cut up into a number of *slices,* each about 200 μm thick. For the purpose of comparison, the thickness of the paper on which this book is printed is about 100 μm. The slice forms the *substrate* of the IC, and is of *p*–type material.

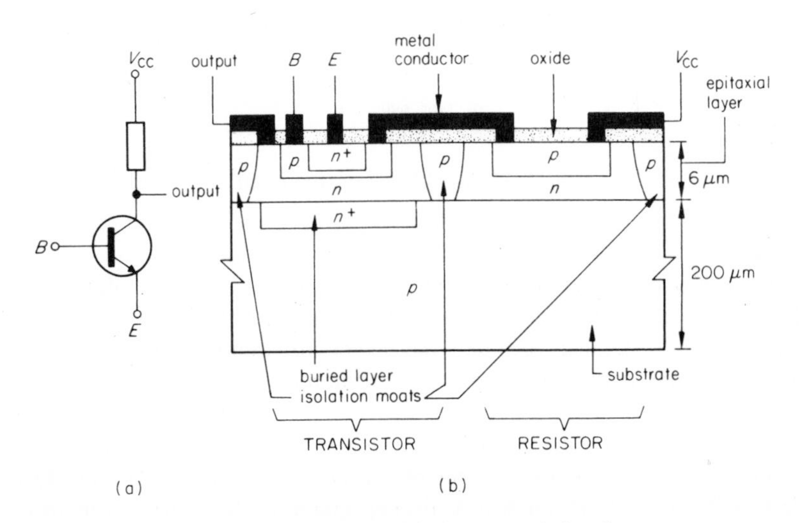

Fig. 5.11 A monolithic integrated circuit

An n^+ layer is then *diffused* in a special furnace into the substrate, and this layer is eventually 'buried' under an n–type *epitaxial layer* (meaning 'arranged upon'). The n^+ symbol simply implies that the impurity doping of the n^+ region is higher than is normal in an n–type material: it results in the n^+ region having an increased conductivity relative to an n–region. The high conductivity buried layer is introduced to reduce the saturation voltage $V_{CE(sat)}$ of the transistor.

An n–type epitaxial layer is then *grown* upon the substrate, and it is in this layer that the complete IC is constructed. The next step is the diffusion of the p–type regions which form the isolation moats between the circuit components. In bipolar IC's, fairly complex isolation techniques need to be adopted to ensure isolation between components. A further p–type diffusion follows to form the base of the transistor and the body of the pull-up resistor. A final n^+ diffusion results in the formation of the transistor emitter. The whole surface is then covered by a thin layer of glass insulation (silicon dioxide), through which 'windows' are etched to allow aluminium connections to be made to the electrodes.

Figure 5.12 shows a simplified cross-sectional diagram of a p–MOST NAND gate, corresponding to the circuit in figure 5.9(b). The diagram shows the relative simplicity of MOS IC construction when compared with bipolar construction. In figure 5.12, the channel length of the pinch-effect transistor is about three times that of the transistors. Also, the channel width of the resistor is about one-quarter that of the transistors.

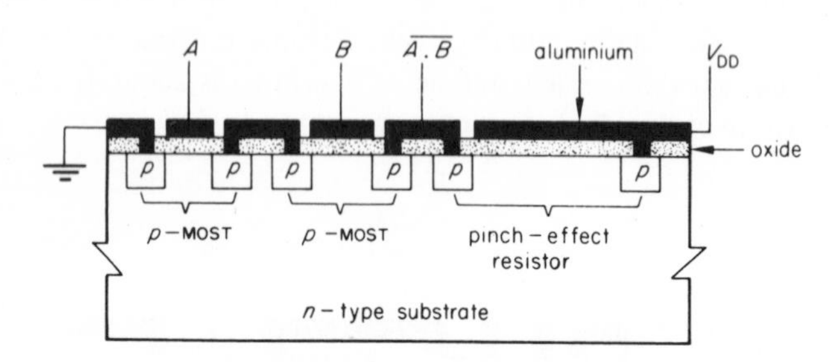

Fig. 5.12 A MOS integrated circuit

Since the gate insulation is extremely thin, its breakdown voltage of about 100 V can be exceeded in normal handling. To prevent the rupturing voltage from being accidentally applied, each gate has a **gate protection diode** connected to it to provide a leakage path which prevents the build-up of stored charge on the gate capacitance.

In the construction of CMOS IC's it is necessary to provide isolation between the p–channel and n–channel devices, thereby reducing the component density in the surface of the IC. One method of achieving isolation is to diffuse a 'tub' of p–type material into the substrate, into which the n–channel devices are introduced.

5.9 LSI, MSI, and SSI

The terms **large scale integrated (LSI) circuit, medium scale integrated (MSI) circuit,** and **small scale integrated (SSI) circuit** are in everyday use, and refer to the number of gates contained in a single IC package. The terms are not always precisely defined, and the following are those in common usage.

LSI contain more than about 100 gates.
MSI contain between about 10 and 100 gates.
SSI contain up to about 10 gates.

5.10 IC packaging

Monolithic IC's are manufactured in three basic types of packages (or **packs**), which are:

1. T05 canisters (or **cans**)
2. flatpacks
3. plastic encapsulated dual-in-line (DIL) packs.

Outline diagrams of the three types are shown in figure 5.13, types 1, 2, and 3 above corresponding to diagrams (a), (b), and (c), respectively, in figure 5.13.

Both flatpacks and T05 cans are hermetically sealed, and can be used over a temperature range of -55° to 125°C, and are generally used where space and weight are at a premium. The plastic DIL pack is cheap to manufacture and is used in the great majority of industrial and commercial applications. With the normal

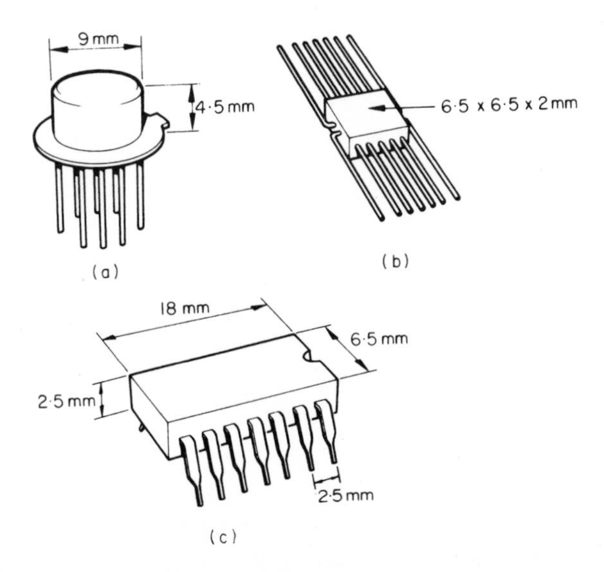

Fig. 5.13 Methods of packaging integrated circuits

range of devices, DIL packs can operate over the temperature range 0 to 70° C, and some types can be used over the wider range given above.

The most popular arrangement is the 14-pin DIL pack, having seven connections or pins per side, the pins being 2.5 mm (0.1 in) apart. Of these, one is required for the power supply and one for the earth line, leaving twelve connections for the input and output of data. A widely used 14-pin DIL pack is the 7400N (or FJH131) TTL quadruple (or **quad**) 2-input NAND pack, the connections to which are shown in figure 5.14.

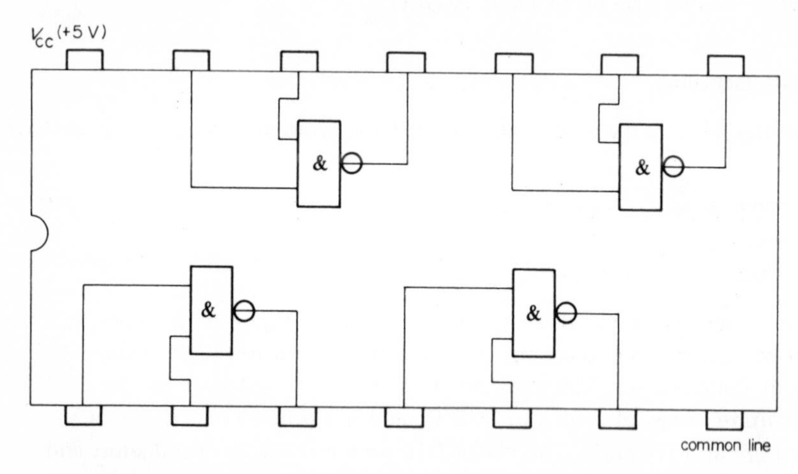

Fig. 5.14 A quadruple (quad) 2-input NAND integrated-circuit package

6 The Algebra of Logic

6.1 The laws of logic

In order to test the truth or otherwise of logical statements we first draw up rules or laws of the processes involved. The principal laws are listed below.

Commutative law

This states that the order in which terms or variables appear in the equation is irrelevant.

$$A + B = B + A$$

$$A \,.\, B = B \,.\, A$$

Associative law

This states that the order in which identical functions are performed in the equation is irrelevant.

$$A + B + C = (A + B) + C = A + (B + C)$$

$$A \,.\, B \,.\, C = (A \,.\, B) \,.\, C = A \,.\, (B \,.\, C)$$

NOTE: Care has been taken in bracketing terms together, since $A + B \,.\, (C + D)$ is not equal to $(A + B) \,.\, (C + D)$, but it is equal to $A + (C + D) \,.\, B$.

Distributive law

This is expressed in two forms.

$$A + (B \,.\, C \,.\, D \,.\, \ldots) = (A + B) \,.\, (A + C) \,.\, (A + D) \,.\, \ldots$$

$$A \,.\, (B + C + D + \ldots) = A \,.\, B + A \,.\, C + A \,.\, D + \ldots$$

The first gives the **product of sums** expression, and the second gives the **sum of products** result.

6.2 Logic theorems

De Morgan's theorem

This states that the logical complement of a function is obtained if we (1) logically invert each term in the expression, and (2) interchange the 'dots' with the 'plusses' and vice versa, as follows.

$$\overline{A \,.\, B \,.\, C} = \bar{A} + \bar{B} + \bar{C}$$

$$\overline{A + B + C} = \bar{A} \,.\, \bar{B} \,.\, \bar{C}$$

NOTE: If we apply the second form above to the DCTL gate in figure 5.3(a), we see that the gate generates the NOR function. This comment also applies to figure 5.9(a).

Other useful theorems

1. $A + 0 = A$
2. $A + 1 = 1$
3. $A + A = A$
4. $A + \bar{A} = 1$
5. $A \,.\, 0 = 0$
6. $A \,.\, 1 = A$
7. $A \,.\, A = A$
8. $A \,.\, \bar{A} = 0$
9. $\bar{\bar{A}} = A$

6.3 Applications of the laws of logic

A fundamental application of the laws and theorems outlined above is to the **analysis** and **simplification** of logical expressions. From such an analysis, it is possible to construct or **synthesise** logic networks; network synthesis is described in chapter 7. The following examples best illustrate the use of the laws and theorems.

Example 6.1

Signals from sensors A, B, and C in a security system must give an indication of the presence of an intruder when the following logical conditions are satisfied.

$$f = A \,.\, B \,.\, C + A \,.\, \bar{B} \,.\, C + A \,.\, B \,.\, \bar{C}$$

Simplify the expression for f.

Solution 6.1

The equation tells us that an alarm signal is initiated when any or all of the three logical groupings on the right-hand side of the equation are satisfied.

Inspecting the terms in the equation, we see that the term $A \,.\, B$ appears in the first and third groups. First then, let us apply the commutative law to collect like terms together, and then apply the distributive law as follows

$$f = A \,.\, B \,.\, C + A \,.\, B \,.\, \bar{C} + A \,.\, \bar{B} \,.\, C \qquad \text{(commutative law)}$$

$$= A \,.\, B \,.\, (C + \bar{C}) + A \,.\, \bar{B} \,.\, C \qquad \text{(distributive law)}$$

We now apply theorem 4 to simplify the first term in the expression.

$$f = A \cdot B \cdot 1 + A \cdot \overline{B} \cdot C \qquad \text{(theorem 4)}$$

Applying theorem 6 we find that

$$f = A \cdot B + A \cdot \overline{B} \cdot C$$

From the second form of the distributive law, we can further say that

$$f = A \cdot (B + \overline{B} \cdot C) \qquad (6.1)$$

This is one possible form of solution. Another form can be deduced from the fact that the term $A \cdot C$ appears in the first and second expressions of the original equation. The equation is then simplified as follows.

$$f = A \cdot C \cdot B + A \cdot C \cdot \overline{B} + A \cdot B \cdot \overline{C} \qquad \text{(commutative law applied twice)}$$

$$= A \cdot C \cdot (B + \overline{B}) + A \cdot B \cdot \overline{C} \qquad \text{(distributive law)}$$

$$= A \cdot C \cdot 1 + A \cdot B \cdot \overline{C} \qquad \text{(theorem 4)}$$

$$= A \cdot C + A \cdot B \cdot \overline{C} \qquad \text{(theorem 6)}$$

$$= A \cdot (C + B \cdot \overline{C}) \qquad (6.2)$$

Although equations (6.1) and (6.2) differ slightly, they both represent the initial equation.

In the above solutions we have used different sections of the term $A \cdot B \cdot C$ in the simplification process. This leads to the idea of another form of solution since, by using theorem 3 in reverse we can say that

$$A \cdot B \cdot C = A \cdot B \cdot C + A \cdot B \cdot C$$

If this is substituted into the original expression we get

$$f = A \cdot B \cdot C + A \cdot B \cdot C + A \cdot \overline{B} \cdot C + A \cdot B \cdot \overline{C} \qquad \text{(theorem 3)}$$

$$= (A \cdot B \cdot C + A \cdot B \cdot \overline{C}) + (A \cdot B \cdot C + A \cdot \overline{B} \cdot C) \qquad \text{(commutative law)}$$

$$= A \cdot B \cdot (C + \overline{C}) + A \cdot C \cdot (B + \overline{B}) \qquad \text{(distributive law)}$$

$$= A \cdot B \cdot 1 + A \cdot C \cdot 1 \qquad \text{(theorem 4)}$$

$$= A \cdot B + A \cdot C \qquad \text{(theorem 6)}$$

$$= A \cdot (B + C) \qquad \text{(distributive law)}$$

Once more we have arrived at a solution which differs from equations (6.1) and (6.2). The solution just obtained is, in fact, the **minimal solution** expressed in terms of a logical algebraic expression. However, the expression 'minimal solution' can be interpreted in several ways, and the *logical* minimal expression is not necessarily the best solution from a circuit viewpoint. For example, we have obtained a number of possible versions of the original equation, and whilst the third version gives the neatest logical solution, it remains to be seen whether it provides the best electrical solution in terms of the number of IC packs required, or of the number and length of interconnections involved, or of the overall speed of operation, or of the overall cost. The solution adopted is often a compromise between these factors.

Example 6.2

Tests on an integrated circuit reveal that it satisfies the truth table given below. Determine the logical function the IC generates, and simplify the expression as far as possible.

inputs			output
A	*B*	*C*	*X*
0	0	0	0
0	0	1	0
0	1	0	1
0	1	1	0

inputs			output
A	*B*	*C*	*X*
1	0	0	0
1	0	1	1
1	1	0	1
1	1	1	1

Solution 6.2

One method of deriving the function generated is to write down the logical equation which gives *all* the conditions which provide a '1' at the output. The first condition occurs in the third row, when

$$A = 0 \text{ (that is, } \bar{A} = 1\text{) and } B = 1 \text{ and } C = 0 \text{ (that is, } \bar{C} = 1\text{)}$$

Thus $X = 1$ when the input condition $\bar{A} \,.\, B \,.\, \bar{C} = 1$ is satisfied. Progressing down the list we see that X is also '1' when the conditions $A \,.\, \bar{B} \,.\, C$, $A \,.\, B \,.\, \bar{C}$, $A \,.\, B \,.\, C$ are satisfied. This is written down in logical form as

$$X = \bar{A} \,.\, B \,.\, \bar{C} + A \,.\, \bar{B} \,.\, C + A \,.\, B \,.\, \bar{C} + A \,.\, B \,.\, C$$

This expression, although complete, has yet to be minimised. To minimise the expressions we group like terms together as far as possible. This we do by grouping the first and third terms, and also the second and fourth terms.

$$\begin{aligned} X &= (\bar{A} \,.\, B \,.\, \bar{C} + A \,.\, B \,.\, \bar{C}) + (A \,.\, \bar{B} \,.\, C + A \,.\, B \,.\, C) \\ &= B \,.\, \bar{C} \,.\, (\bar{A} + A) + A \,.\, C \,.\, (\bar{B} + B) \\ &= B \,.\, \bar{C} + A \,.\, C \qquad (6.3) \end{aligned}$$

Equation (6.3) is one form of minimal expression which satisfies the truth table.

An alternative approach is to say that **all** the conditions defined by 1's in the truth table are simply **NOT the 0's.** As the 1's in the truth table define X, then the zeros in the truth table define the function $\overline{X}$, so that

$$\overline{X} = \overline{A} \,.\, \overline{B} \,.\, \overline{C} + \overline{A} \,.\, \overline{B} \,.\, C + \overline{A} \,.\, B \,.\, C + A \,.\, \overline{B} \,.\, \overline{C}$$

By grouping the first term in the expression for $\overline{X}$ with the fourth term, and the second term with the third term, we see that

$$\begin{aligned} \overline{X} &= (\overline{A} \,.\, \overline{B} \,.\, \overline{C} + A \,.\, \overline{B} \,.\, \overline{C}) + (\overline{A} \,.\, \overline{B} \,.\, C + \overline{A} \,.\, B \,.\, C) \\ &= \overline{B} \,.\, \overline{C} \,.\, (\overline{A} + A) + \overline{A} \,.\, C \,.\, (\overline{B} + B) \\ &= \overline{B} \,.\, \overline{C} + \overline{A} \,.\, C \end{aligned} \tag{6.4}$$

Now, from theorem 9, $\overline{\overline{X}} = X$ and, if we complement (that is, NOT) both sides of equation (6.4) we arrive at an equation for X of

$$X = \overline{\overline{X}} = \overline{\overline{B} \,.\, \overline{C} + \overline{A} \,.\, C} \tag{6.5}$$

Equation (6.5) is another minimal form of the original equation. Although equations (6.4) and (6.5) differ in appearance they are, in fact, equivalent to one another as we shall show in the following.

In equation (6.5) let $M = \overline{B} \,.\, \overline{C}$ and $N = \overline{A} \,.\, C$. Applying De Morgan's theorem to equation (6.5) we see that

$$X = \overline{M + N} = \overline{M} \,.\, \overline{N} = (\overline{\overline{B} \,.\, \overline{C}}) \,.\, (\overline{\overline{A} \,.\, C})$$

Applying De Morgan's theorem once more gives

$$\begin{aligned} X &= (B + C) \,.\, (A + \overline{C}) && \text{(De Morgan)} \\ &= A \,.\, B + B \,.\, \overline{C} + A \,.\, C + C \,.\, \overline{C} && \text{(distributive law)} \\ &= A \,.\, B + B \,.\, \overline{C} + A \,.\, C + 0 && \text{(theorem 8)} \\ &= A \,.\, B + B \,.\, \overline{C} + A \,.\, C && \text{(theorem 1)} \\ &= A \,.\, B \,.\, (C + \overline{C}) + (B \,.\, \overline{C} + B \,.\, \overline{C}) + (A \,.\, C + A \,.\, C) && \text{(theorems 3 and 4)} \\ &= (A \,.\, B \,.\, C + A \,.\, B \,.\, \overline{C} + B \,.\, \overline{C} + A \,.\, C) + (B \,.\, \overline{C} + A \,.\, C) \\ &= (A \,.\, C \,.\, (B + 1) + B \,.\, \overline{C} \,.\, (A + 1)) + (B \,.\, \overline{C} + A \,.\, C) \\ &= (A \,.\, C + B \,.\, \overline{C}) + (B \,.\, \overline{C} + A \,.\, C) && \text{(theorems 2 and 6)} \\ &= A \,.\, C + B \,.\, \overline{C} && \text{(theorem 3)} \end{aligned}$$

which is equivalent to equation (6.3)

Example 6.3

In designing a logic network which controls part of a computer switching system, the following function must be satisfied.

$$f = W \,.\, (\overline{\overline{X} + \overline{W} \,.\, (Y + \overline{X} \,.\, \overline{Y})}) \tag{6.6}$$

Simplify the expression.

Solution 6.3

The first step is to write the equation in the form $f = W \,.\, D$

where

$$D = \overline{\overline{X} + \overline{W} \,.\, (Y + \overline{X} \,.\, \overline{Y})} \,.$$

Expanding the expression under the 'bar' we get

$$D = \overline{\overline{X} + \overline{W} \,.\, Y + \overline{W} \,.\, \overline{X} \,.\, \overline{Y}}$$

Further, if we let $E = \overline{W} \,.\, Y$, and $F = \overline{W} \,.\, \overline{X} \,.\, \overline{Y}$,

then

$$D = \overline{\overline{X} + E + F} = X \,.\, \overline{E} \,.\, \overline{F} \qquad \text{(De Morgan)}$$

$$= X \,.\, (\overline{\overline{W} \,.\, Y}) \,.\, (\overline{\overline{W} \,.\, \overline{X} \,.\, \overline{Y}})$$

Now, by De Morgan's theorem

$$\overline{\overline{W} \,.\, Y} = W + \overline{Y}$$

and

$$\overline{\overline{W} \,.\, \overline{X} \,.\, \overline{Y}} = W + X + Y$$

Hence

$$D = X \,.\, (W + \overline{Y}) \,.\, (W + X + Y)$$

$$= (W \,.\, X + X \,.\, \overline{Y}) \,.\, (W + X + Y)$$

$$= W \,.\, X + W \,.\, X + W \,.\, X \,.\, Y + W \,.\, X \,.\, \overline{Y} + X \,.\, \overline{Y} + X \,.\, Y \,.\, \overline{Y}$$

$$= W \,.\, X + X \,.\, \overline{Y}$$

Now

$$f = W \,.\, D = W \,.\, (W \,.\, X + X \,.\, \overline{Y})$$

$$= W \,.\, (W \,.\, X \,.\, (Y + \overline{Y}) + X \,.\, \overline{Y} \,.\, (W + \overline{W}))$$

$$= W \,.\, (W \,.\, X \,.\, Y + W \,.\, X \,.\, \overline{Y} + \overline{W} \,.\, X \,.\, \overline{Y})$$

$$= W \,.\, X \,.\, Y + W \,.\, X \,.\, \overline{Y} + W \,.\, \overline{W} \,.\, X \,.\, \overline{Y}$$

$$= W \,.\, X \,.\, (Y + \overline{Y}) = W \,.\, X$$

That is, the complex logical function described by equation (6.6) is simply equivalent to W AND X, that is, input Y is **redundant** and plays no part in the operation of the system.

7 Logic Networks

7.1 Combinational logic and sequential logic networks

Logic networks fall into two broad classes, namely **combinational logic networks and sequential logic networks.** Combinational logic networks include the networks dealt with so far, and which generate an output signal when a given combination or combinations of input signals exist.

The output from sequential logic systems depend on the sequence of events which have already occurred in the circuit, and include such systems as counters. In this chapter we deal with the design of combinational logic systems.

7.2 The design of logic networks from truth tables

As we have seen in chapter 6, we can deduce the logical equation which completely defines a truth table, and in the following we shall see how to design a logic system from the equation. Consider the equation in table 7.1.

Table 7.1

inputs			output
A	*B*	*C*	*X*
0	0	0	0
0	0	1	1
0	1	0	0
0	1	1	0
1	0	0	1
1	0	1	1
1	1	0	0
1	1	1	0

Writing down the expression which defines the 1's in the table gives

$$X = \bar{A} \,.\, \bar{B} \,.\, C + A \,.\, \bar{B} \,.\, \bar{C} + A \,.\, \bar{B} \,.\, C \qquad (7.1)$$

If the sensors which energise the logic system detect the state of the variables *A*, *B*, and C, then we need three NOT gates to generate the functions $\bar{A}$, $\bar{B}$, and $\bar{C}$ required in equation (7.1). These gates are shown in figure 7.1(a).

The function $\bar{A} . \bar{B} . C$ is generated by a 3-input AND gate, its inputs being energised by signals $\bar{A}$, $\bar{B}$, and C in the manner shown in figure 7.1(b). The output from this gate is designated the letter L. In equation (7.1) the terms $A . \bar{B} . \bar{C}$ (= M) and $A . \bar{B} . C$ (= N) are generated as shown in figure 7.1(c).

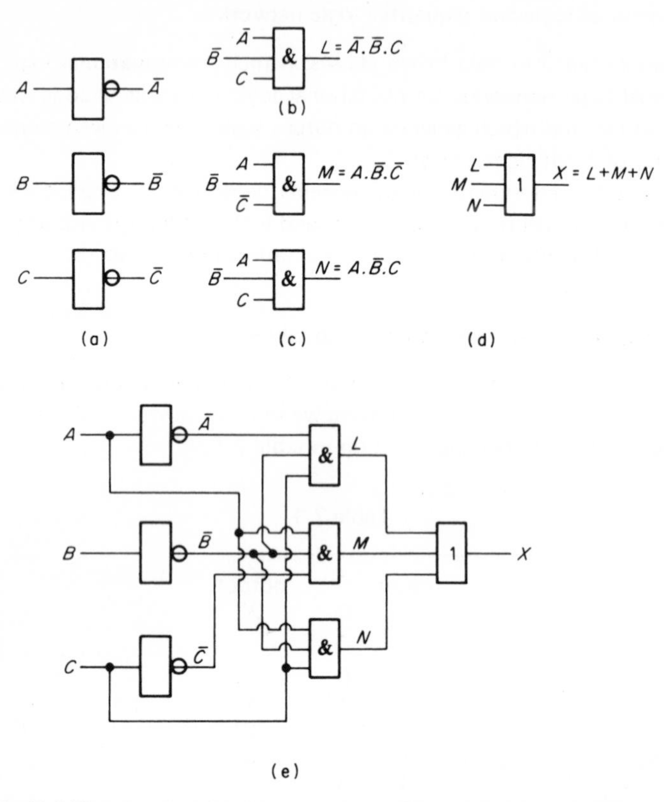

Fig. 7.1 The elements generating the basic functions in equation 7.1

Finally the function X is generated by OR-gating the terms L, M, and N as shown in figure 7.1(d). The completed block diagram is shown in figure 7.1(e), in which the individual sections are linked together.

Using the techniques outlined in chapter 6, equation (7.1) can be simplified to

$$X = A . \bar{B} + C . \bar{B} \tag{7.2}$$

$$= \bar{B} . (A + C) \tag{7.3}$$

The block diagram of the logic network which satisfies equation (7.2) is shown in figure 7.2, and requires four gates. The logic block diagram corresponding to

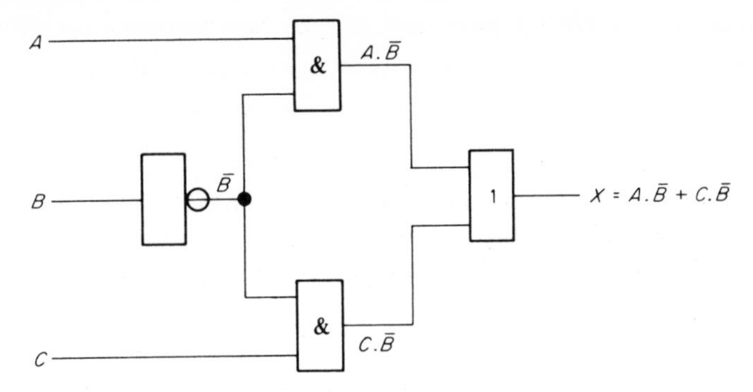

Fig. 7.2 A logic block diagram satisfying equation 7.2

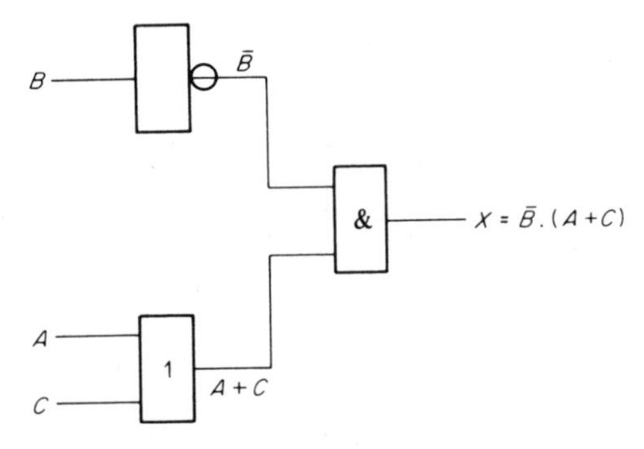

Fig. 7.3 A logic block diagram satisfying equation 7.3

equation (7.3) is shown in figure 7.3, and only three gates are required in this case.

An alternative approach is to develop an equation which defines NOT the 0's in the truth table. Such an equation is

$$\overline{X} = \overline{A}\,.\,\overline{B}\,.\,\overline{C} + \overline{A}\,.\,B\,.\,\overline{C} + \overline{A}\,.\,B\,.\,C + A\,.\,B\,.\,\overline{C} + A\,.\,B\,.\,C$$

Applying De Morgan's theorem to the above equation gives

$$X = (A + B + C)\,.\,(A + \overline{B} + C)\,.\,(A\,.\,\overline{B}\,.\,\overline{C})\,.\,(\overline{A} + \overline{B} + C)\,.\,(\overline{A} + \overline{B} + \overline{C})$$

Although this is a complex logical statement, it is well suited for direct implementation in the form of a NOR network (see section 7.4).

7.3 NAND networks

TTL NAND gates are perhaps the most popular elements in use today, and it is possible to generate all the functions so far described by using combinations of these gates. In this section of the book we shall consider how to develop various functions using only NAND gates, and also see how complete networks can be constructed.

The NOT function

As shown earlier, if we use a 1-input NAND gate in the manner of figure 7.4(a), then its output is logic '0' when the input is logic '1'. Also, when $A = 0$, then the output is '1'. That is, a 1-input NAND gate functions as a NOT gate.

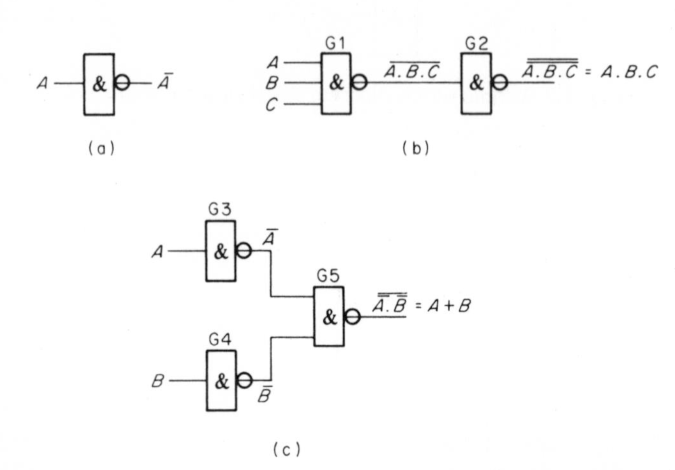

Fig. 7.4 Methods of generating (a) the NOT function, (b) the AND function and (c) the OR function using NAND gates only

The AND function

Let us consider the operation of the circuit in figure 7.4(b). Gate G1 functions as a 3-input NAND gate, whose output is $\overline{A \,.\, B \,.\, C}$. Gate G2 acts as a NOT gate which logically inverts the signal at the output of G1, that is, the output is

$$\overline{(\overline{A \,.\, B \,.\, C})} = A \,.\, B \,.\, C$$

which is the AND function of the inputs.

The OR function

In the circuit in figure 7.4(c), gates G3 and G4 act as invertors whose outputs are $\overline{A}$ and $\overline{B}$, respectively. Gate G5 operates as a 2-input NAND gate whose inputs are $\overline{A}$ and $\overline{B}$, to give an output of

$$\overline{\overline{A} \,.\, \overline{B}} = A + B$$

which is the OR function of the inputs.

The NOR function

Since NOR = NOT OR, this function is generated by driving the input of a NOT circuit of the type in figure 7.4(a) by an OR circuit of the type in figure 7.4(c).

Basic minimisation techniques for NAND networks

Basic types of NAND networks may require a large number of gates, but it may be possible in many cases to reduce the number of gates required by the application of simple rules.

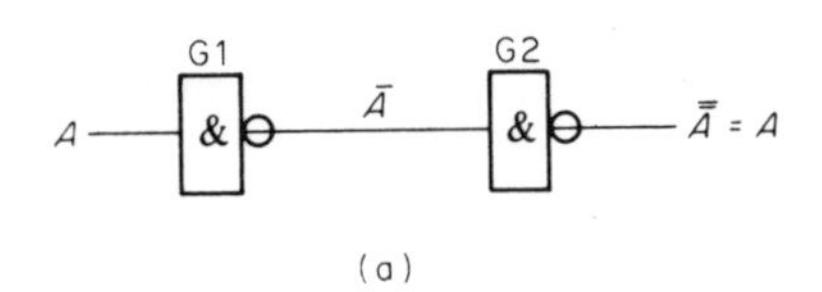

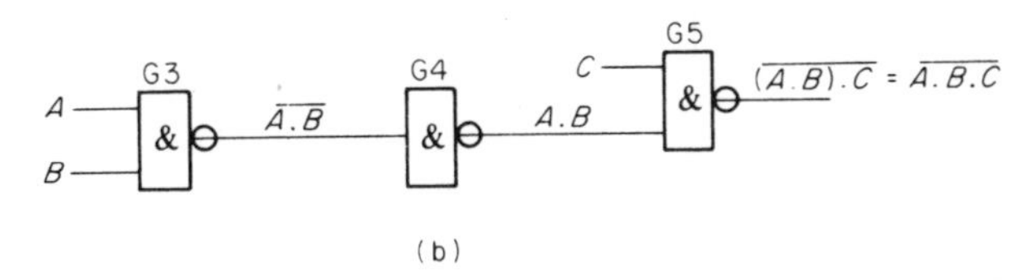

Fig. 7.5 Basic minimisation techniques for use in NAND networks

If two 1-input NAND gates are cascaded in the manner shown in figure 7.5(a), the output from G1 is $\overline{A}$, and that from G2 is $\overline{\overline{A}}$ which, by theorem 9 in chapter 6, is equal to A. That is, **both gates are redundant and can be replaced by a single wire connecting the input to the output.**

In the following we show that the network in figure 7.5(b) can be replaced by a 3-input NAND gate. The output from G3 is $\overline{A \, . \, B}$, which is logically inverted by G4, whose output is $A \, . \, B$. The function generated by the 2-input gate G5 is $\overline{(A \, . \, B) \, . \, C} = \overline{A \, . \, B \, . \, C}$. This function would be generated by a single 3-input NAND gate whose inputs are energised independently by signals A, B, and C.

The results of figure 7.5(b) can also be used in the reverse order to advantage in some cases. Suppose that we have a number of IC packages containing only 2-input NAND gates (for example, 7400N quad 2-input NAND packs), and we need to generate the function $\overline{A \, . \, B \, . \, C}$. Using the circuit in figure 7.5(b), we can use three 2-input gates to replace a single 3-input gate. In some instances the use of this technique can lead to a high utilization of IC packs.

Replacing AND–OR networks by NAND networks

When we write down the logical equation defining all the 1's in a truth table, we obtain an equation which is in the *sum of products* form, that is, in an AND–OR type of equation. An example of this type was given in equation (7.1) of section 7.2. To illustrate the basic principles, we shall consider the design of a NAND network which generates the function $A \,.\, B + C \,.\, D$. Using AND and OR gates, the network required is shown in figure 7.6(a). To replace figure 7.6(a) by NAND elements, we first replace each AND element by its NAND equivalent (see figure 7.4(b)), and replace the OR element by the NAND equivalent as in figure 7.4(c). The resulting NAND network is shown in figure 7.6(b). Now, using the minimisation technique in figure 7.5(a), we remove gates G2 and G5 in figure 7.6(b) and replace them by a single wire, and similarly we also replace gates G4 and G6. This leaves only gates G1, G2, and G7 in figure 7.6(b), and the simplified NAND network is shown in figure 7.6(c).

Comparing circuits (a) and (c) in figure 7.6, we see that the AND–OR network is

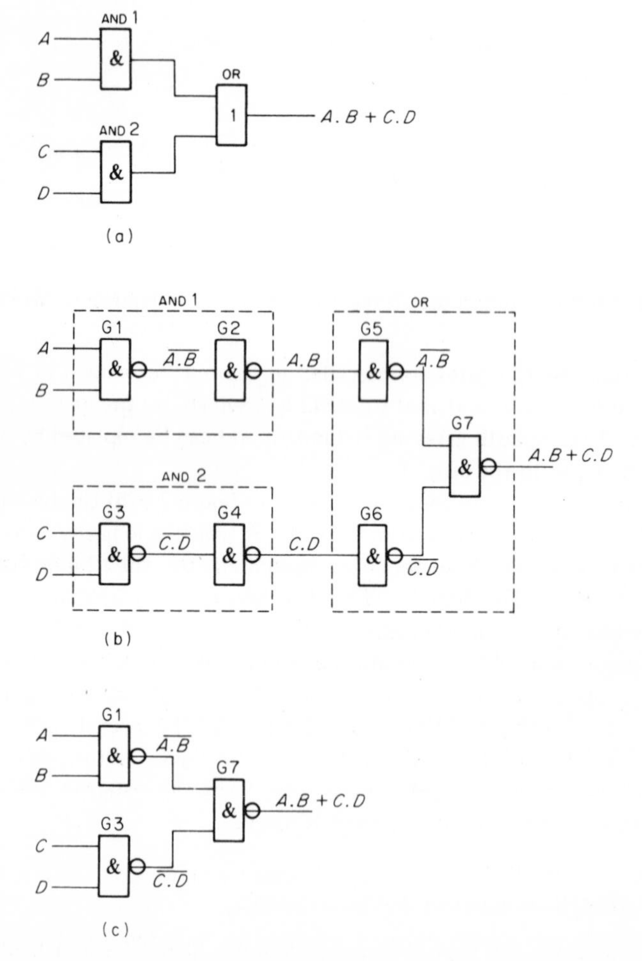

Fig. 7.6 Generating the function $f = A \,.\, B + C \,.\, D$ using NAND gates

replaced by an equivalent all-NAND network. This technique is applicable to any AND–OR type of network, and readers may like to show that if we replace *all* of the gates in figure 7.1(e) by NAND gates, then the output of the resulting all-NAND network is given by equation (7.1).

7.4 NOR networks

As with NAND gates, NOR gates can be used to generate the basic logic functions. The block diagrams associated with these networks are considered in the following sections.

The NOT function

We saw when first dealing with the NOR gate that when any input is energised by a logic '1', it causes the output to be logic '0'. Using a 1-input NOR gate, as in figure 7.7(a), when the input is logic '0', the output is logic '1'. That is, a 1-input NOR gate acts as a NOT gate.

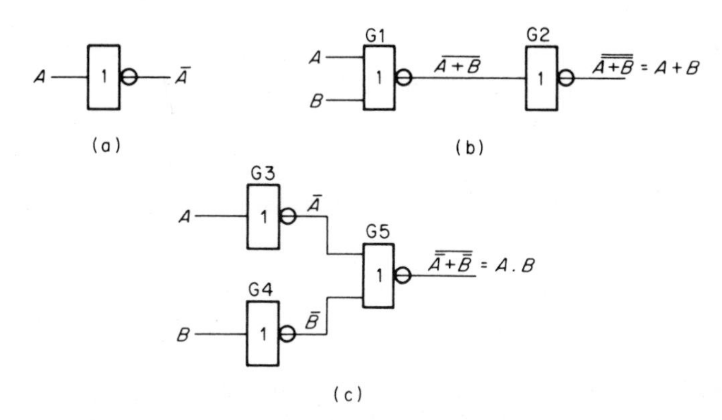

Fig. 7.7 Methods of generating (a) the NOT function, (b) the OR function and (c) the AND function using NOR gates only

The OR function

In figure 7.7(b), the output from G1 is $\overline{A+B}$, and that from G2 is $\overline{\overline{A+B}} = A+B$, that is, it is the OR function of the inputs.

The AND function

For two input signals, this function is developed by figure 7.7(c). Gates G3 and G4 act as invertors to give signals $\bar{A}$ and $\bar{B}$ respectively. The output from G5 is $\overline{\bar{A}+\bar{B}}$ which, by De Morgan's theorem, is $A \,.\, B$, that is, it is the AND function of the input signals.

The NAND function

Since NAND = NOT AND, we can develop this function by using an AND gate of the type in figure 7.7(c) to drive the input of a NOT gate similar to that in figure 7.7(a), the output of which is the NAND function of the inputs to the combined circuit.

Basic minimisation techniques for NOR networks

The minimisation techniques developed here are similar to those used earlier with NAND gates.

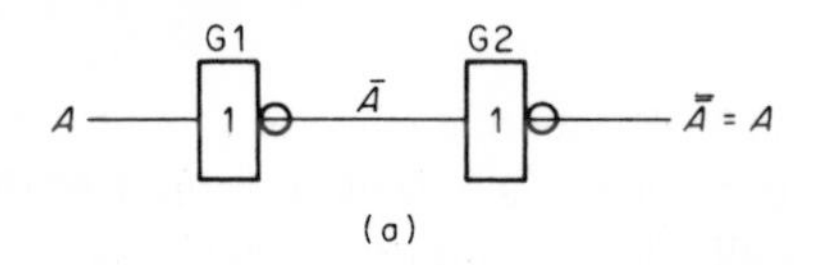

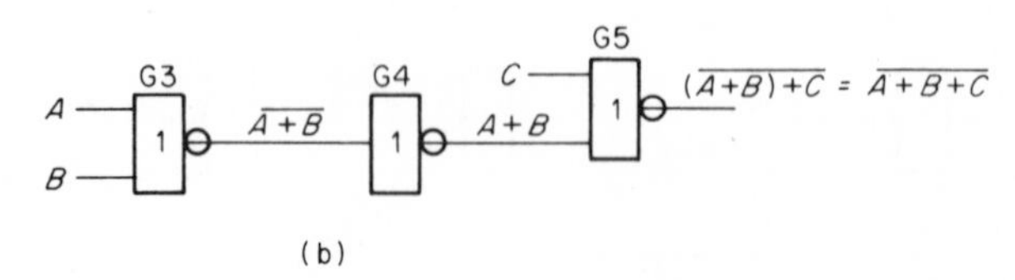

Fig. 7.8 Basic minimisation techniques for use in NOR networks

In figure 7.8(a), the double inversion of the single input signal results in both gates being redundant, and they can be replaced by a wire linking the input to the output.

In figure 7.8(b), G3 and G4 form an OR gate (see also figure 7.7(b)), so that the output from G5 is $\overline{(A + B) + C} = \overline{A + B + C}$. That is, the complete circuit in figure 7.8(b) can be replaced by a single 3-input NOR gate. This technique can also be used in reverse to allow us to design NOR networks using gates with a limited fan-in to generate functions requiring a large number of inputs.

Replacing OR–AND networks with NOR networks

It was shown in section 7.2 that the logical equation which defines all the 0's in a truth table can be written in the *logical product-of-sums* form, that is, in an OR–AND form such as $(A + B) \,.\, (C + D)$.

Let us consider a network which generates the function $(A + B) \,.\, (C + D)$. The basic OR–AND network is shown in figure 7.9(a) and, using the circuits in figure 7.7, it is implemented in NOR form in figure 7.9(b). Readers will note that gates G2 and G5 form a cascaded pair of NOR gates which have a single input and

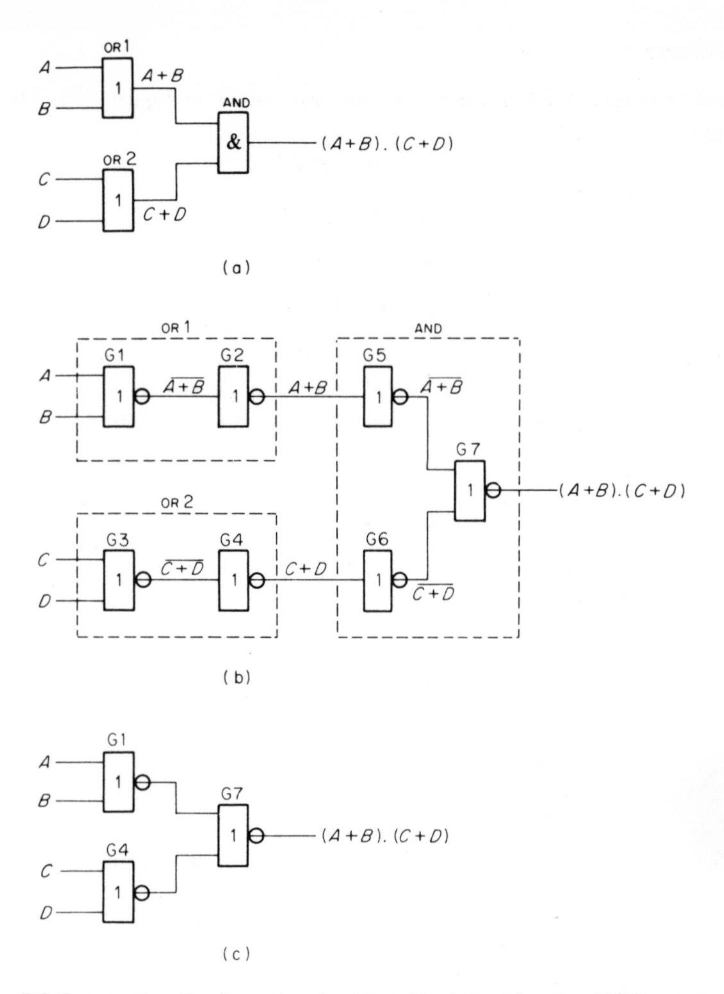

Fig. 7.9 Generating the function $f = (A + B) \, . \, (C + D)$ using NOR gates only

are therefore redundant, as are gates G4 and G6. This leaves gates G1, G3 and G7 to generate the required function. The resulting NOR network is shown in figure 7.9(c).

7.5 Wired-OR networks

The wired-OR connection (see also section 4.15) can be used with either NAND or NOR gates provided that they have pull-up resistors in their output circuits. It was shown in section 4.15 that the function generated by this connection could be regarded as the AND function of the individual outputs from the gates. To illustrate the effect on NOR and NAND networks, we shall consider each separately.

NOR networks

Two gates connected by a wired-OR link are shown in figure 7.10. The logical expression for output f is

$$f = (\overline{A + B}) \,.\, (\overline{C + D})$$

$$= (\bar{A} \,.\, \bar{B}) \,.\, (\bar{C} \,.\, \bar{D})$$

$$= \bar{A} \,.\, \bar{B} \,.\, \bar{C} \,.\, \bar{D}$$

$$= \overline{A + B + C + D}$$

That is, when NOR gates are used in wired-OR networks the resulting function generated is the NOR function of all the input signals. Hence the wired-OR connection can be used as a means of increasing the fan-in of NOR networks.

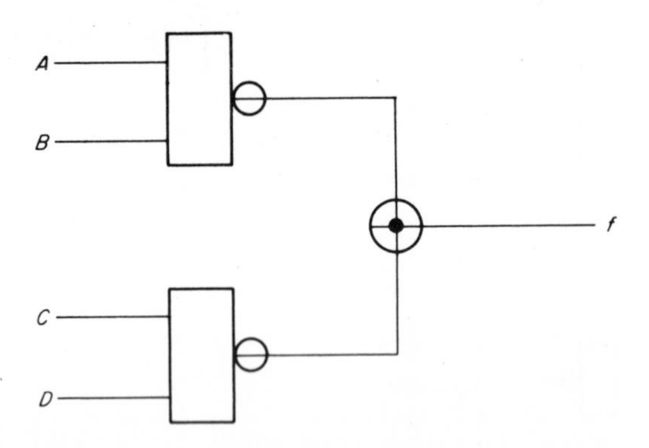

Fig.7.10 The wired-OR connection of NOR gates

NAND networks

If we use the network in figure 7.10 once more, but this time using NAND gates, the function generated is

$$f = (\overline{A \,.\, B}) \,.\, (\overline{C \,.\, D})$$

$$= \overline{A \,.\, B + C \,.\, D}$$

7.6 The exclusive-OR function

Gates generating the exclusive-OR function are widely used in logic circuits for operations including addition, subtraction, multiplication, division, and binary number comparison. The truth table defining the function is given in table 7.2.

The first three rows of table 7.2 are identical to those of the 2-variable OR function (see table 2.2). Table 7.2 differs from that of the OR function only in the final line since, in table 7.2, when $A = B = 1$ the output is logic '0'.

Table 7.2

inputs A	B	output S
0	0	0
0	1	1
1	0	1
1	1	0

In this part of the book we shall consider the design of networks which generate the exclusive-OR function, applications of the circuit being introduced at appropriate points throughout the book. Inspecting table 7.2, we see that output S has the value '1' when $A\,.\,\overline{B} = 1$ or when $\overline{A}\,.\,B = 1$, that is

$$S = A\,.\,\overline{B} + \overline{A}\,.\,B \tag{7.4}$$

This function is implemented in figure 7.11(a) using AND, OR, and NOT gates. The term $A\,.\overline{B}$ is generated by G1, $\overline{A}\,.\,B$ is generated by G2, and G3 forms the final output. Two symbols used to represent exclusive-OR gates are shown in figure 7.11(b).

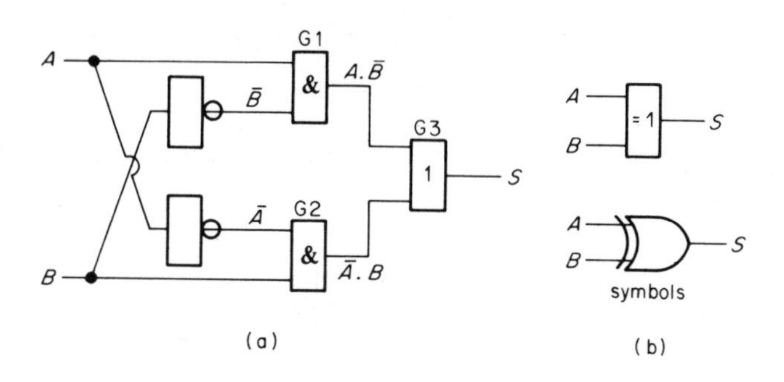

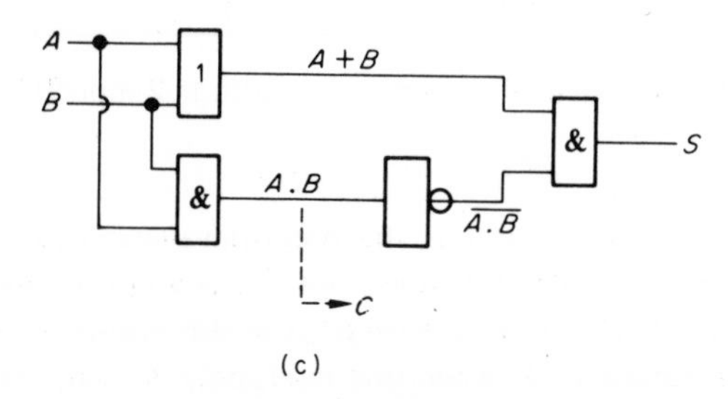

Fig. 7.11 The exclusive-OR function

An exclusive-OR circuit which requires only four gates can be developed by applying our knowledge of boolean algebra to the equation for S as follows.

$$S = A\,.\,\bar{B} + \bar{A}\,.\,B + 0 + 0 \qquad \text{(theorem 1)}$$

$$= A\,.\,\bar{B} + \bar{A}\,.\,B + A\,.\,\bar{A} + B\,.\,\bar{B} \qquad \text{(theorem 8)}$$

$$= (A + B)\,.\,(\bar{A} + \bar{B}) \qquad (7.5)$$

$$= (A + B)\,.\,(\overline{A\,.\,B}) \qquad (7.6)$$

The block diagram of a circuit which generates equation (7.6) is shown in figure 7.11(c). The circuit also generates at point C the function $A\,.\,B$. This additional output is particularly useful in arithmetic circuits, and is discussed further in chapter 9.

Equation (7.5) can also be written in the form

$$S = A\,.\,(\bar{A} + \bar{B}) + B\,.\,(\bar{A} + \bar{B})$$

$$= A\,.\,(\overline{A\,.\,B}) + B\,.\,(\overline{A\,.\,B})$$

This function is generated by the NAND network in figure 7.12, since the output from this circuit is

$$S = \overline{\overline{(A\,.\,(\overline{A\,.\,B}))}\,.\,\overline{(B\,.\,(\overline{A\,.\,B}))}}$$

$$= A\,.\,(\overline{A\,.\,B}) + B\,.\,(\overline{A\,.\,B})$$

The circuit in figure 7.12 can be constructed using a single quad 2-input NAND IC package (see section 5.10).

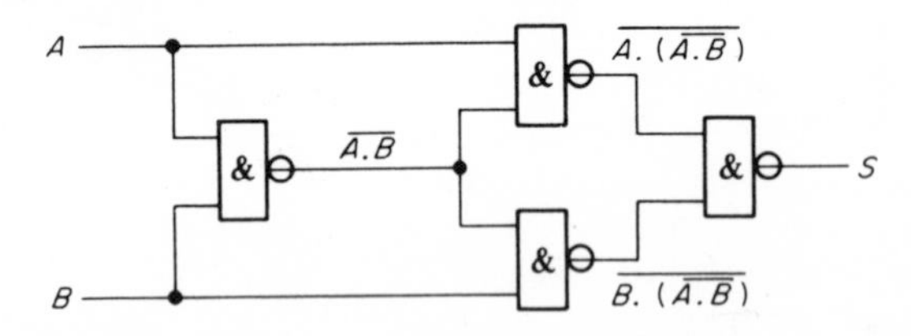

Fig. 7.12 Generating the exclusive-OR function using four 2-input NAND gates

The exclusive-OR gate is also known as the **non-equivalence** gate since from table 7.2, the output is logic '1' when the inputs are logically not equivalent to one another, that is, when $A = 1$, $B = 0$ or vice versa. It is also known as a **modulo-2** addition circuit since the output is '1' when one input only is energised, otherwise the output is zero.

7.7 Karnaugh maps

So far we have attempted to minimise networks using boolean algebra only. In some instances the reason for steps taken during the minimisation procedure are not always obvious, and it often takes a little time to obtain a satisfactory solution. An alternative minimisation technique uses what is known as a Karnaugh map named after the scientist who devised it. The Karnaugh map is simply a method of plotting or mapping all the conditions given in the truth table.

Single variable map

A single variable can have one of two possible operating states; if the variable is designated the symbol A, then it has either the logical value '1' ($A = 1$), or it has the value '0' ($\bar{A} = 1$). In the case of a single variable, the Karnaugh map contains two equal divisions, known as **cells,** as shown in figure 7.13(a), one half of the map representing the state of A and the other half representing $\bar{A}$. If we wish to represent the condition $A = 1$, we do so by drawing the map in figure 7.13(b). The $\bar{A}$ cell contains zero, since, when $A = 1$, variable $\bar{A} = 0$. We represent the condition $\bar{A} = 1$ (that is, $A = 0$) by the map in figure 7.13(c).

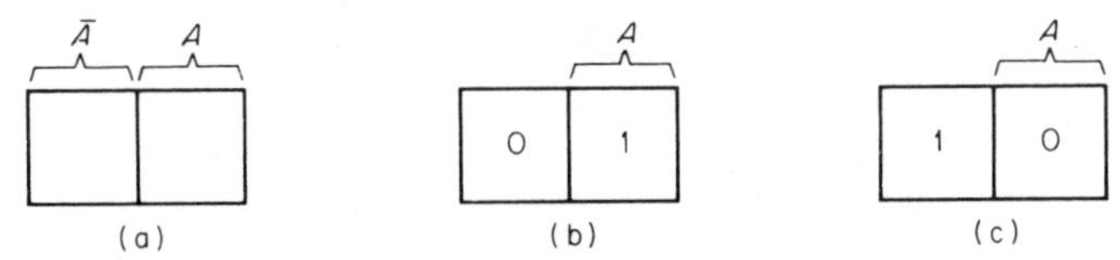

Fig. 7.13 Karnaugh maps for a single variable

A 2-variable map

Each variable associated with a problem has two possible states, so that when there are two variables we have four possible combinations of the variables. Hence the 2-variable maps in figure 7.14 have four cells to represent these states. There are two popular methods of drawing Karnaugh maps, both being shown in figure 7.14, and each having its advantages. Let us consider the map in figure 7.14(a).

As before, each variable must be represented by one-half of the total number of cells, and in this case we allow variable A to represent the cells in the right-hand

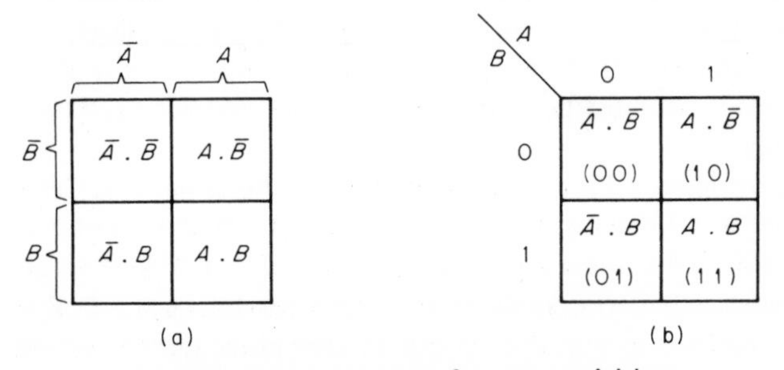

Fig. 7.14 Karnaugh maps for two variables

half of the map. The two cells in the left-hand column represent $\bar{A}$. The two lower cells represent the variable B, and the two upper cells represent $\bar{B}$. Each cell in the map is defined by the **intersection** or **union** of the variables, much as a point on a geographical map can be defined by horizontal and vertical references. Thus the cell in the lower right-hand corner is the intersection of the variables A and B, and is shown as the cell $A \, . \, B$. The cell directly above it is located by the intersection of variables A and (NOT B), and is shown as the cell $A \, . \, \bar{B}$. The other cells on the map are defined in much the same way.

In the map in figure 7.14(b), we represent variable A by placing a '1' at the head of the column representing A, and a '0' at the head of the $\bar{A}$ column. Similarly a '1' placed at the left-hand end of the lower row of cells shows that that row represents variable B, and a '0' at the left-hand end of the upper row indicates that it represents $\bar{B}$. The binary grouping written inside the cells (that is, 00, 10, 01, 11) represents the values of the **input** variables associated with the cells. For example, when $A = 0$ and $B = 1$ the cell concerned is defined by the binary group 01, that is, the cell $\bar{A} \, . \, B$. We shall now illustrate how the Karnaugh map is used in problems involving logic.

Let us consider how we map the function given in table 7.3.

Table 7.3

inputs A	B	output f	cell defined by the inputs
0	0	0	$\bar{A} \, . \, \bar{B}$
0	1	0	$\bar{A} \, . \, B$
1	0	0	$A \, . \, \bar{B}$
1	1	1	$A \, . \, B$

The Karnaugh map associated with table 7.3 is illustrated in figure 7.15(a), and is formed as follows. Considering the table a row at a time, the input conditions corresponding to the first row are $A = 0$, $B = 0$, that is, we are concerned with the cell $\bar{A} \, . \, \bar{B}$ on the map. We place in this cell the value of the function f which is the output from the network. Since $f = 0$ in this case, we place a '0' in the upper left-hand cell (cell $\bar{A} \, . \, \bar{B}$) of the map in figure 7.15(a). Similarly, cells $\bar{A} \, . \, B$ and $A \, . \, \bar{B}$ have 0's placed in them. Since a '1' appears in the final row of the f column of the truth table, a '1' must be written in the lower right-hand cell (cell $A \, . \, B$) of the Karnaugh map.

The map in figure 7.15(b) is formed in much the same way. Cell 00 (the upper left-hand cell) is defined by the input conditions $A = 0, B = 0$ and in the truth table we note that $f = 0$ for this state. Accordingly we write a '0' in that cell. Cell 01 (the lower left-hand cell) corresponds to the input conditions $A = 0$, $B = 1$, for which $f = 0$; a '0' is written in that cell. In cell 11 (the lower right-hand cell) we record a '1' since $f = 1$ when $A = 1$ and $B = 1$.

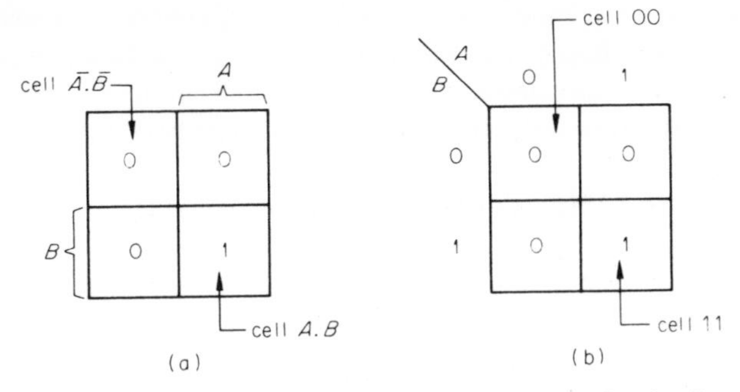

Fig. 7.15 Karnaugh maps for the function $f = A \,.\, B$

Now suppose we have 1's in more than one cell, as shown in figure 7.16(a), in which the cells defined by 1's are $A \,.\, B$ and $A \,.\, \bar{B}$. A logical network which generates an output corresponding to the map in figure 7.16(a) provides a '1' output either when $A \,.\, B = 1$ OR when $A \,.\, \bar{B} = 1$, that is

$$f = A \,.\, B + A \,.\, \bar{B}$$

Applying the rules of boolean algebra to the above equation, we simplify it as follows

$$f = A \,.\, (B + \bar{B}) = A \,.\, 1 = A$$

That is, the map in figure 7.16(a) defines the function $f = A$. We can derive the same result from the Karnaugh map by grouping together **adjacent cells** which

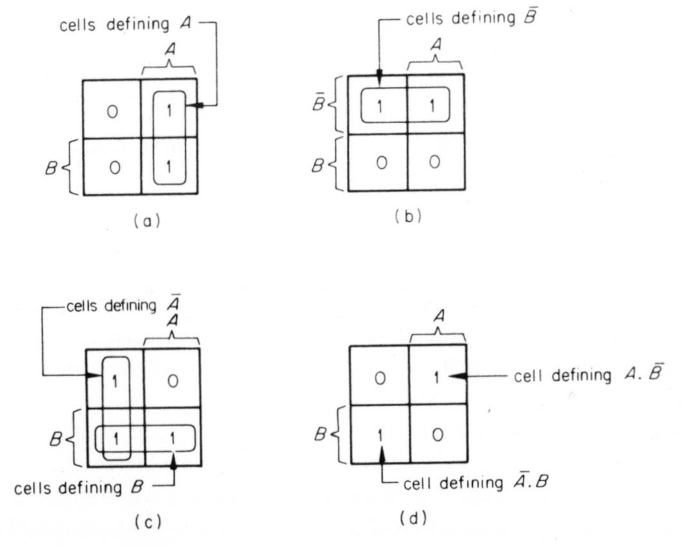

Fig. 7.16 Grouping adjacent cells on a two-variable map

contain 1's in the manner shown in figure 7.16(a). *Adjacent cells are defined as cells which differ in their binary representation by only one* **binary digit (bit).** For example, cell $A \, . \, B$ is represented by the binary input group 11, and cell $A \, . \, \overline{B}$ by the group 10; these differ only in the value of the right-hand bit. More will be said about this later.

The map in figure 7.16(b) is that of the logical equation

$$f = \overline{A} \, . \, \overline{B} + A \, . \, \overline{B}$$

which, by boolean algebra, can be reduced to

$$f = \overline{B} \, . \, (\overline{A} + A) = \overline{B}$$

Once more, we can group two adjacent cells on the Karnaugh map in order to simplify the defining equation, as shown in figure 7.16(b).

The map in figure 7.16(c) includes 1's in three cells, and the logical equation satisfied by the map is

$$f = A \, . \, B + \overline{A} \, . \, B + \overline{A} \, . \, \overline{B} \tag{7.7}$$

This equation can be simplified to $f = \overline{A} + B$ by boolean algebra, but the steps involved are not obvious at first sight. However, if we group the cells containing 1's in the manner shown in figure 7.16(c), we see that the horizontal group of cells includes *all* cells defined by variable B and the vertical group of cells includes *all* cells defined by $\overline{A}$, hence

$$f = \overline{A} + B$$

This introduces us to the concept that any '1' on the Karnaugh map can be used several times, a fact which is proved in the following. From theorem 3 (chapter 6), we see that

$$\overline{A} \, . \, B + \overline{A} \, . \, B = \overline{A} \, . \, B$$

that is, the term $\overline{A} \, . \, B$ (or any other term for that matter) may be repeated as *many times* as we like (the process is not limited simply to its use twice) once it has appeared in our equation. Thus, if we repeat $\overline{A} \, . \, B$ twice in equation (7.7), we get

$$\begin{aligned} f &= A \, . \, B + (\overline{A} \, . \, B + \overline{A} \, . \, B) + \overline{A} \, . \, \overline{B} \\ &= (A \, . \, B + \overline{A} \, . \, B) + (\overline{A} \, . \, B + \overline{A} \, . \, \overline{B}) \qquad (7.8) \\ &= B \, . \, (A + \overline{A}) + \overline{A} \, . \, (B + \overline{B}) = B + \overline{A} \end{aligned}$$

The first bracketed pair of terms in equation (7.8) corresponds to the lower pair of cells in figure 7.16(c), whilst the second bracketed pair of terms corresponds to the pair of cells in the left-hand column of figure 7.16(c).

An example of a function which cannot be simplified is mapped in figure 7.16(d). This represents the function $A \, . \, \overline{B} + \overline{A} \, . \, B$, that is, the exclusive-OR function, the map containing 1's in cells which are not adjacent to one another. These cells are defined by input conditions 10 and 01 and, as explained above, are not adjacent since their binary representations differ in more than one bit.

A 3-variable map

A 3-variable map is shown in figure 7.17 and, since there are $2^3 = 8$ possible combinations of the input variables, there are eight cells in the Karnaugh map. Once again, each variable must be represented in one-half of the total number of cells, so that variable A is represented in four cells and $\bar{A}$ in the remaining four. Variable B is also represented in four cells, two of which are associated with variable A and two with $\bar{A}$; variable $\bar{B}$ also links with two cells containing A and two containing $\bar{A}$. In this way, all possible combinations of variables A and B are obtained. Similarly variable C links with both A and B so that all possible combinations of A, B, and C are generated.

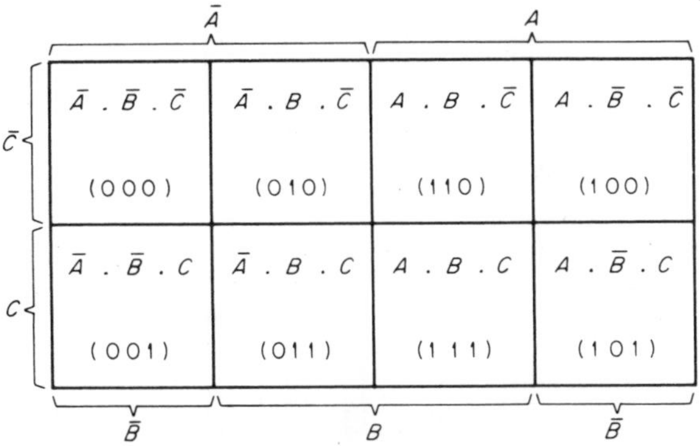

	$\bar{A}$		A	
$\bar{C}$	$\bar{A} . \bar{B} . \bar{C}$ (000)	$\bar{A} . B . \bar{C}$ (010)	$A . B . \bar{C}$ (110)	$A . \bar{B} . \bar{C}$ (100)
C	$\bar{A} . \bar{B} . C$ (001)	$\bar{A} . B . C$ (011)	$A . B . C$ (111)	$A . \bar{B} . C$ (101)
	$\bar{B}$	B		$\bar{B}$

Fig. 7.17 A three-variable Karnaugh map

The binary code groups locating the cells are shown in figure 7.17 and, once more, we note that adjacent cells differ by only one bit in their respective code groups. An interesting feature of this map is the fact that the cells at the extreme left-hand side and the extreme right-hand side of each row are *adjacent cells* according to our definition. For example, the cells at the extremes of the upper row are the cells 000 and 100, which differ only in the left-hand digit; cells 001 and 101 in the bottom row differ only in the left-hand digit. Since we have *side-to-side adjacency,* we can 'bend' the map to form a continuous cylinder with the ends forming the 'seam' of the cylinder.

Examples of function mapping on 3-variable maps are shown in figure 7.18 The function mapped in figure 7.18(a) contains 1's in non-adjacent cells and cannot, therefore, be simplified. The expression defined by this map is

$$f = A . \bar{B} . C + \bar{A} . B . \bar{C}$$

Figure 7.18(b) contains two pairs of adjacent cells, Let us first deal with the lower row. The cells grouped here are

$$A . B . C + \bar{A} . B . C = (A + \bar{A}) . B . C = B . C$$

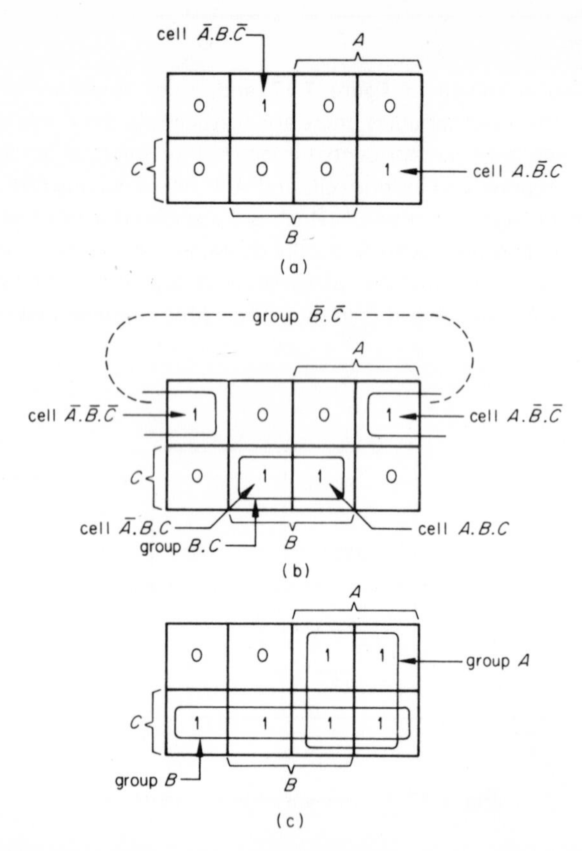

Fig. 7.18 Illustrating the meaning of adjacent cells on a three-variable map

The equation defining this group can readily be deduced from the Karnaugh map by noting that the two centre cells in the bottom row represent the *total* number of cells covered by the *intersection* of variables B and C. Similarly, the two cells containing 1's in the upper row have side-to-side adjacency, and form the total number of cells represented by the intersection of variables $\bar{B}$ and $\bar{C}$. This fact is verified in the following

$$A \,.\, \bar{B} \,.\, \bar{C} + \bar{A} \,.\, \bar{B} \,.\, \bar{C} = (A + \bar{A}) \,.\, \bar{B} \,.\, \bar{C} = \bar{B} \,.\, \bar{C}$$

Hence the function mapped in figure 7.18(b) is

$$f = B \,.\, C + \bar{B} \,.\, \bar{C}$$

Readers will note that that the function A is missing from the above expression, hence variable A is **redundant** in this problem.

Two methods of grouping four adjacent cells are shown in figure 7.18(c). Let us consider the right-hand group, which define the logical expression

$$A\,.\,B\,.\,C + A\,.\,B\,.\,\bar{C} + A\,.\,\bar{B}\,.\,\bar{C} + A\,.\,\bar{B}\,.\,C$$
$$= A\,.\,(B\,.\,C + B\,.\,\bar{C} + \bar{B}\,.\,\bar{C} + \bar{B}\,.\,C)$$
$$= A\,.\,(B\,.\,(C + \bar{C}) + \bar{B}\,.\,(\bar{C} + C))$$
$$= A\,.\,(B + \bar{B}) = A$$

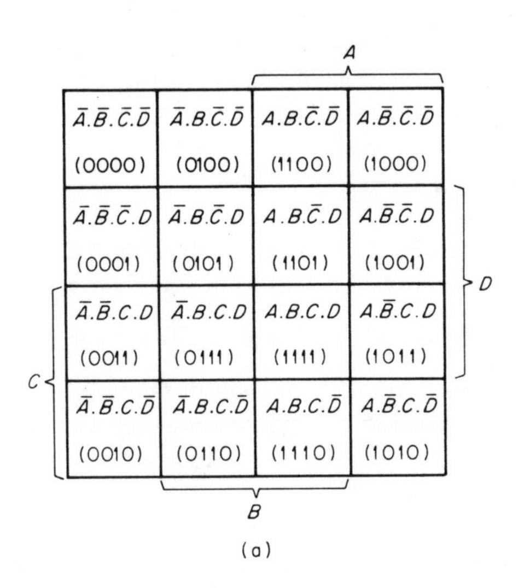

(a)

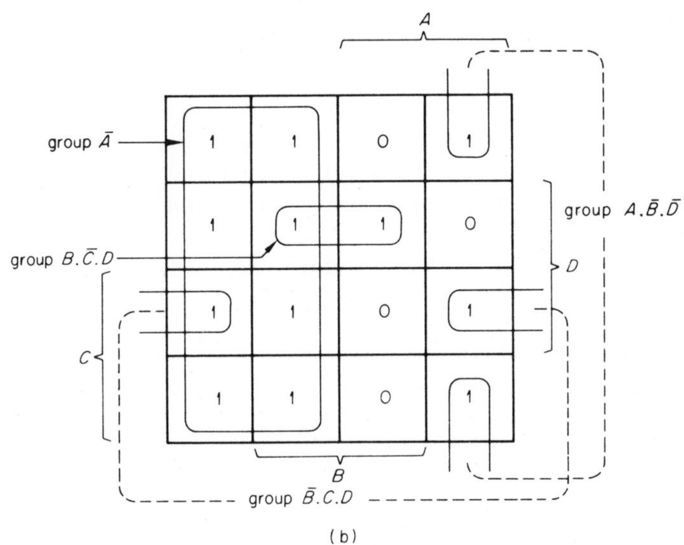

(b)

Fig. 7.19 (a) A four-variable map and (b) the map of the function

$$f = \bar{A} + B\,.\,\bar{C}\,.\,D + \bar{B}\,.\,C\,.\,D + A\,.\,\bar{B}\,.\,\bar{D}$$

This result can be obtained directly from the Karnaugh map, since the four cells listed above are the total number of cells defined by variable A on the map. It can also be seen that the four cells in the lower row represent variable C, this fact can be verified by a similar argument to that above. The logical function defined by the map in figure 7.18(c) is, therefore

$$f = A + C$$

It can now be noted that **the larger the number of cells grouped on the Karnaugh map, the simpler is the final logical expression.**

If three adjacent cells are to be grouped together on the map, then they may be grouped as two separate pairs in the manner outlined in figure 7.16(c).

A 4-variable map

A 4-variable map containing $2^4 = 16$ cells is shown in figure 7.19(a); an example of a 4-variable map is shown in figure 7.19(b) in which the function mapped is

$$f = \bar{A} + B \,.\, \bar{C} \,.\, D + \bar{B} \,.\, C \,.\, D + A \,.\, \bar{B} \,.\, \bar{D}$$

The eight cells which are grouped on the left of figure 7.19(b) collectively represent the term $\bar{A}$, while the term $B \,.\, \bar{C} \,.\, D$ is represented by the pair of cells grouped together in the centre of the map. An example of side-to-side adjacency is illustrated by the group $\bar{B} \,.\, C \,.\, D$.

4-variable maps illustrate another facet of adjacency on Karnaugh maps, which is *top-to-bottom adjacency*. When we inspect the code groups at the head and foot of each column in figure 7.19(a), we find that they satisfy the conditions associated with adjacent cells. That is to say, they differ in only one bit, and we can bend the top of the map over to touch the bottom to form a cylinder. We have already shown that the ends are adjacent, so that the Karnaugh map can be regarded as being a toroid or continuous cylinder, in the manner of an inner tube of a tyre In this way, we see that the four corner cells in figure 7.19(a) form an adjacent group $\bar{B} \,.\, \bar{D}$ on the toroidal map. An example of top-to-bottom adjacency is illustrated in the group $A \,.\, \bar{B} \,.\, \bar{D}$.

The mapping technique can be extended to deal with more than four variables.

8 Memory Circuits

8.1 Static and dynamic memories

Sequential logic systems include such networks as counters and shift registers, and require the use of memory elements or **flip–flops** to record the state of the circuit at a particular instant of time.

Semiconductor memory circuits fall into two broad categories, namely **static memories** and **dynamic memories.** Static memories retain the stored information indefinitely so long as the power supply is maintained; both bipolar and MOS devices are used in static memory elements. The most popular types of static memory elements are **set–reset (S–R) flip–flops, J–K flip–flops, trigger (T) flip–flops,** and **D flip–flops.** The J–K element is the most versatile of these, since it can be used to generate all other types of memory function.

Dynamic memories depend for their operation on the ability of the gate capacitance of MOS devices to retain their charge for relatively long periods of time (long, that is, compared with the time for a complete cycle of operations within the system, which may only be a few hundred μs in a computer system). Dynamic memories are dealt with in section 8.10.

8.2 The S–R flip–flop

A basic S–R flip–flop using cross-connected NOR gates is shown in figure 8.1(a). The output from the circuit is designated the letter Q, its logical complement being available at terminal $\bar{Q}$. A logical '1' signal applied to the S-line causes Q to be set to the logic '1' level irrespective of its previous state. We may therefore regard the S-line as the line which allows us to **set** the output to the '1' state. At the same instant of time, output $\bar{Q}$ becomes '0'. The application of a signal to the R-line causes Q to be **reset** to '0' (that is, $\bar{Q}$ is set to '1'). A detailed description of the operation of the circuit follows.

Let us assume that $Q = 0$ initially (that is, $\bar{Q} = 1$), and that the signals applied to both input lines are zero. This state will be seen to be a stable operating condition, since the logic '1' output from G2 which is fed back to G1 holds the output of G1 at '0'. Both inputs to G2 are 0's, so that its output is '1'.

The application of a logic '1' signal to the S-line causes the output of G2 to fall to zero and, since this is fed back to the input of G1, then the output of G1 rises to logic '1'. This is the second stable operating condition with Q having been set to '1'.

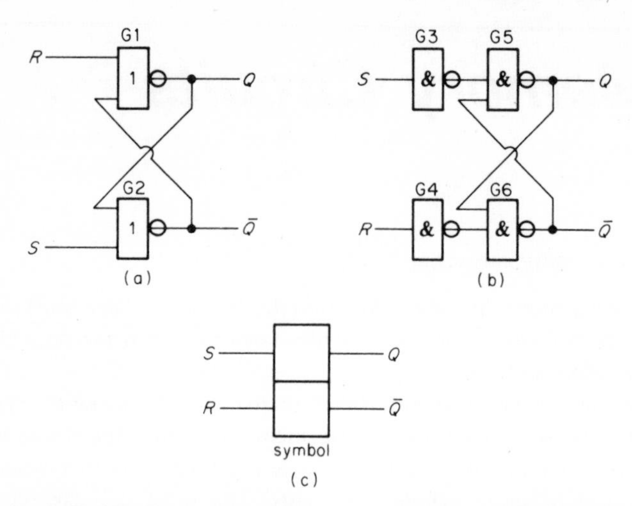

Fig. 8.1 An S–R flip–flop or bistable using (a) NOR gates and (b) NAND gates. (c) Circuit symbol of the S–R flip–flop

The signal applied to the S-line need only be applied momentarily, since the feedback connections between the gates cause the flip–flop to memorise the last instruction. The application of further pulses to the S-line has no further effect on the state of the circuit, since it has already been set to the '1' state.

It can be shown by a similar argument that an impulsive '1' signal applied to the R-line (with input $S = 0$) causes Q to be reset to zero.

The application of '1' signals to both S and R lines simultaneously causes both outputs to fall to zero. The final states of the outputs when both input signals are removed simultaneously is indeterminate, as this depends on the relative switching speed of the two gates. **This operating condition is avoided in practice.**

Since we are concerned with a sequence of events which changes with time, we define the operation of the network in terms of its **sequential truth table**, which is given in table 8.1.

Table 8.1

S	R	Q_{n+1}	comment
0	0	Q_n	no change
0	1	0	reset
1	0	1	set
1	1	?	indeterminate

In table 8.1, Q_n is the state of the flip–flop output after n operations have occurred, and Q_{n+1} is the state of Q after $n + 1$ operations. Let us assume that, in each case, we have completed n operations, and that the next step of input conditions are the $(n + 1)$th set of input signals.

As stated above, when $S = R = 0$ the output remains unchanged whatever its previous value (it may have either been '1' or '0'). That is, $Q_{n+1} = Q_n$. The second row of the truth table corresponds to the reset condition ($R = 1, S = 0$), causing Q to be zero after the operation. The third row of the truth table is the set operation ($S = 1, R = 0$), which causes Q to be set to the '1' state. When $S = R = 1$, in the case of a NOR memory, both outputs are zero. This is described as an indeterminate or 'don't know' condition since, as outlined above, the final state of the output when the input signals are reduced to zero is indeterminate.

A NAND S–R flip-flop is shown in figure 8.1(b), and includes invertors G3 and G4 in the input lines. The flip-flop itself comprises gates G5 and G6. The explanation of the operation of this circuit is left as an exercise for readers to test their skill on. The S–R flip-flop (either NOR or NAND) is represented by the symbol in figure 8.1(c).

8.3 The gated S–R flip-flop

It is often convenient to control all the operations in a system by means of a common synchronising pulse or **clock pulse.** By this means it is possible to **gate** or to **clock** signals into the flip-flop at a precise time. One method of doing this is shown in figure 8.2.

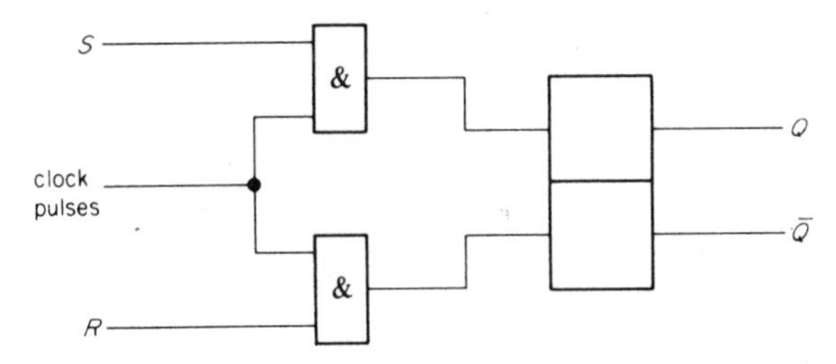

Fig. 8.2 A gated S–R flip-flop

In the case of a NAND flip-flop of the type in figure 8.1(b), no additional gates are required as the clock signal would be applied simultaneously to gates G3 and G4 in figure 8.1(b).

8.4 Contact bounce eliminators

Electrical switches are often used as a means of applying signals to digital systems, but all conventional switches generate electrical noise which results from 'contact bounce'. If the system to which the signal is applied is a counter, then the system counts each noise pulse as though it were a genuine ON-OFF signal. Two methods of eliminating the effects of contact bounce are shown in figure 8.3.

With an S–R flip-flop using NOR gates the circuit in figure 8.3(a) is adopted since, in NOR networks, the gates are switched by logic '1' signals. When switch X

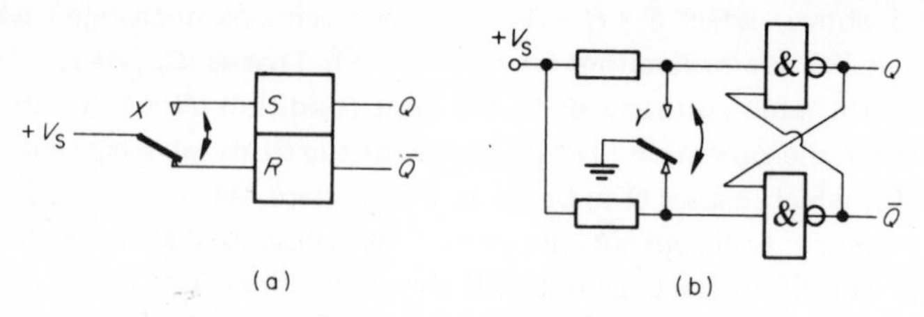

Fig. 8.3 Contact bounce elimination circuits

is moved to the S position, output Q instantaneously becomes '1' and does not alter no matter how many times the switch contacts bounce. Output Q is used as the effective output from the switch.

In NAND circuits, logic '0' signals control the switching operation. The circuit in figure 8.3(b) is one which is frequently used for contact bounce elimination in NAND circuits. Switching the zero potential point to the input of one of the gates causes the output of that gate to be '1'.

8.5 Master–slave flip–flops

In the quest for higher operating speeds various types of flip–flop were developed, and the basis of modern flip–flops is the master–slave circuit.

The underlying principle of the master–slave flip–flop is shown in figure 8.4(a), which contains two S–R flip–flops connected by synchronously-operated switches

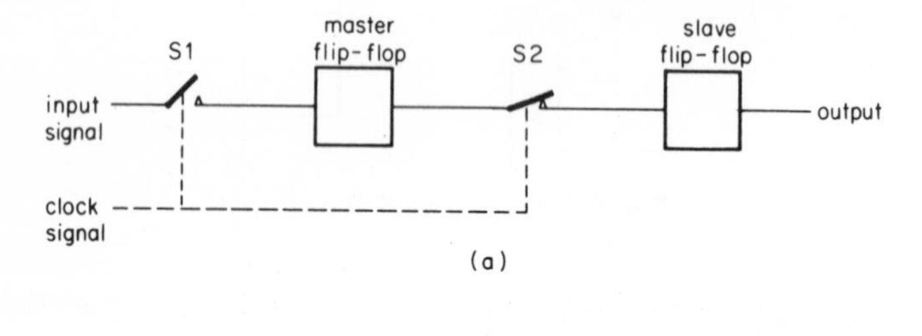

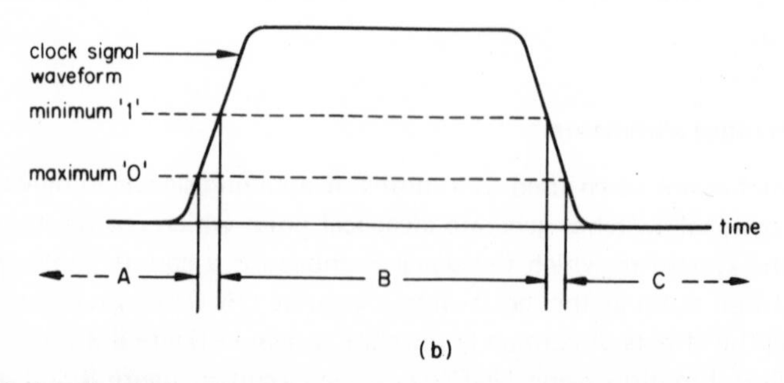

Fig. 8.4 (a) General principles of the master–slave flip–flop and (b) its timing sequence

S1 and S2. The switches are connected so that when S1 is open then S2 is closed, and vice versa. When the clock signal is '0', S1 is open and S2 is closed, so that the information stored in the master is transmitted to the slave. The waveform in figure 8.4(b) is that of the clock signal applied to the flip–flop, and the conditions outlined above occur in period A.

When the clock signal rises to the '1' level, corresponding to period B in figure 8.4(b), S1 closes and S2 opens. At this instant of time, new data is transferred into the master, whilst the slave retains the previous data. When the clock signal falls to zero once more, period C in figure 8.4(b), S1 opens and S2 closes. This isolates the master from the input and connects it to the slave, and the new data is transferred into the slave. Thus, **the new input data is gated to the output of the flip–flop at the trailing edge of the clock pulse.**

A block diagram of a **master–slave S–R flip–flop** is shown in figure 8.5, in which G1 and G2 are equivalent to S1 in figure 8.4(a), and G3 and G4 are equivalent to S2. The invertor G5 provides the correct phase relationship between the two switches. The master and slave flip–flops in figure 8.5 are circuits of the type described in section 8.2.

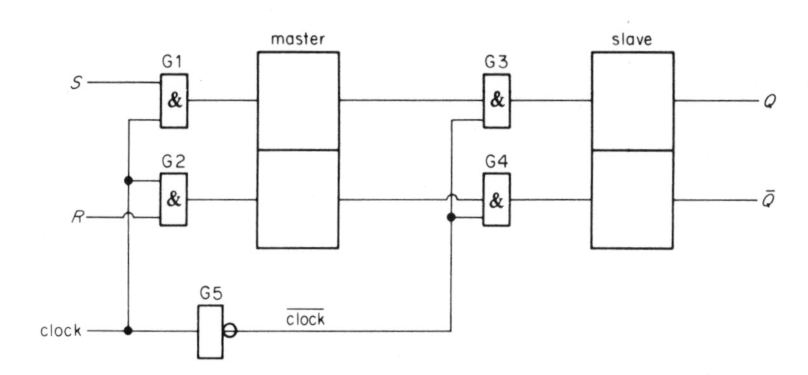

Fig. 8.5 One form of master–slave S–R flip–flop

TTL monolithic IC flip–flops can operate at clock frequencies of up to about 35 MHz.

8.6 The master–slave J–K flip–flop

A basic form of master–slave J–K flip–flop is shown in figure 8.6. It is similar in structure to the master–slave S–R flip–flop, with the exception that the outputs are fed back as shown in figure 8.6. Other input lines are added to the circuit and, although they do not alter the operation of the circuit, they increase the versatility of the flip–flop. The addition of gates G6 and G7 allow the output either to be set to the '1' state or to be reset to the '0' state. In some flip–flops the reset line is known as the **clear** input. As before, gates G1 and G2 are equivalent to S1 in

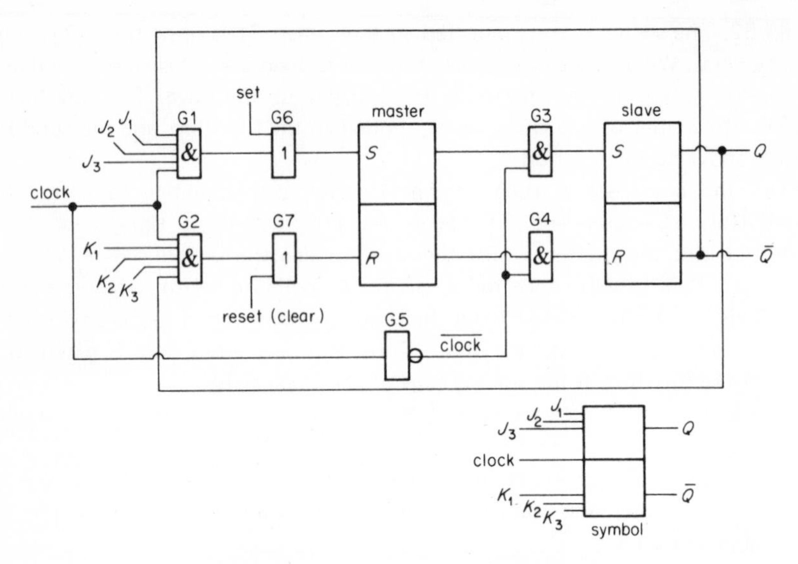

Fig. 8.6 A J–K master–slave flip–flop

figure 8.4(a), and G3 and G4 are equivalent to S2 in the same figure. The sequential truth table for figure 8.6 is given in table 8.2.

Table 8.2

J	K	Q_{n+1}	comments
0	0	Q_n	no change } S–R operation
0	1	0	reset } S–R operation
1	0	1	set } S–R operation
1	1	$\bar{Q}_n$	toggle or trigger

Before going on to describe the operation of the circuit, we shall discuss the implications of the truth table. Comparing tables 8.1 and 8.2, we see that if we regard the J-input and the K-input lines as being equivalent to the S-input and R-input, respectively, of the S–R flip–flop then, for the first three groups of input conditions, the S–R and J–K truth tables are equivalent to one another. That is, the S–R flip–flop may be replaced by the J–K flip–flop.

Also, when $J = K = 1$, the circuit acts as a **toggle flip–flop** or **trigger flip–flop** whose output changes state at the end of *each* clock pulse. That is, if $Q = 0$ initially, then at the end of the first clock pulse the output changes to '1', at the end of the second clock pulse it changes to '0' again, and the process is repeated indefinitely so long as $J = K = 1$. Flip–flops operating in the toggle or trigger mode are widely used in **asynchronous counting systems** (see chapter 10). The operation of the circuit in figure 8.6 is now described.

When $J = K = 0$, both G1 and G2 are inhibited and no signals can be applied to the master flip–flop. As a result, the output remains unchanged and $Q_{n+1} = Q_n$.

The operation for $J = 1$, $K = 0$ is considered in two parts, namely the operation when the initial value of Q is '0' and when it is '1'. Let us consider the case when $Q = 0$. In this case $\bar{Q} = 1$, and when the clock signal is applied it causes G1 to be opened to allow the '1' signal to be applied to the set line of the master flip–flop. Gate G2 is inhibited not only by the '0' on the K-line but also by the '0' fed back from the Q output. When the clock signal finally falls to zero, the '1' stored in the Q output of the master stage is gated into the slave section of the flip–flop, causing Q to change from '0' to '1'.

In the case when Q is initially '1' ($\bar{Q} = 0$), the $\bar{Q}$ signal inhibits the operation of G1, and the states of the master and slave remain unaltered and the output remains at '1'.

The circuit operation for $J = 0$, $K = 1$ is similar to that described above, but for J read K, for K read J, for G1 read G2, for G2 read G1, for Q read $\bar{Q}$, and for $\bar{Q}$ read Q.

For the input condition $J = K = 1$, then **either** G1 **or** G2 is opened when the clock signal is applied, the gate selected being dependent upon the signals fed back from the output. Suppose, initially, that $Q = 0$ and $\bar{Q} = 1$. These signals result in G1 being opened and G2 being inhibited, so that a '1' is fed from the J-line into the master flip–flop. At the end of the first clock pulse, this '1' signal is applied to the set-line of the slave, causing Q to change from '0' to '1'. The feedback conditions have now changed so that a '1' is fed back to G2 and a '0' is applied to G1, thereby opening G2 and inhibiting G1. The next clock pulse causes a '1' to be applied to the reset-line of the master flip-flop. This, at the end of the clock pulse, causes output Q to change from '1' to '0', thereby setting up conditions for a '1' to be fed to Q during the following clock pulse period.

During the period of time that the clock signal is '0', gates G3 and G4 are opened, so that the output can either be set or reset (cleared) by the application of a signal to the set or reset-lines connected to G6 and G7, respectively.

8.7 The trigger (T) flip–flop

A flip–flop with a trigger or toggle operation is constructed merely by connecting the J and K-inputs to a logic '1' signal as shown in figure 8.7(a). The circuit then functions in the manner described in the penultimate paragraph of section 8.6. A symbol frequently used for the T flip–flop is shown in figure 8.7(b), the T-line being synonymous with the C-line in (a).

J–K flip–flops also function as T flip–flops when the J and K-inputs are disconnected. This removes the necessity in some cases of energising the J and K-inputs by a logic '1' signal. An application using this connection is described in section 10.3.

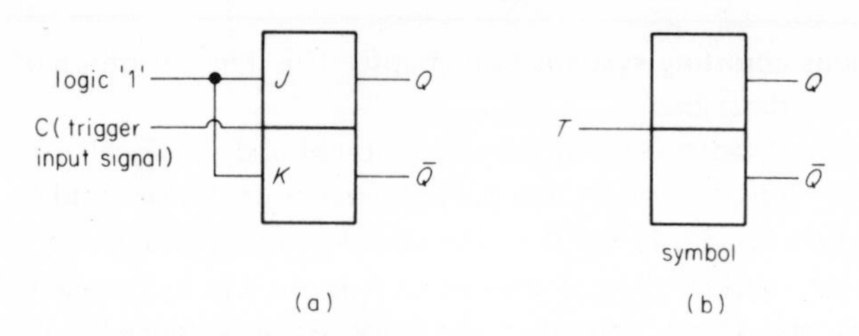

Fig. 8.7 A T (trigger) flip–flop constructed from a J–K flip–flop

8.8 The D master–slave flip–flop

A basic block diagram of a D-type master–slave flip–flop is shown in figure 8.8, and consists of a J–K flip–flop whose K-input is driven via an invertor. The result of adding this invertor is that the signals applied to the J and K-inputs are *always* complementary, and the truth table of the D flip–flop corresponds to the second and third lines of table 8.2, which are listed in table 8.3.

Table 8.3

D	Q_{n+1}	comment
0	0	reset
1	1	set

The D flip–flop, known as a **delay** flip–flop or as a **data** flip–flop, delays the transmission of data between the input and output by a time interval equal to one clock pulse. It is widely used as a data latching buffer element between a counting circuit and a digital read-out device.

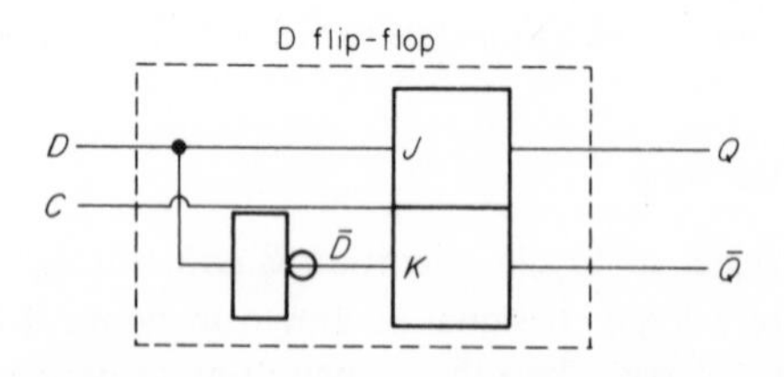

Fig. 8.8 A D (delay) flip–flop constructed from a J–K flip–flop

The D flip–flop has the advantage over the J–K flip–flop in that it only has one input line, resulting in simple interconnections between elements.

Toggle or trigger operation is obtained using the D flip–flop by feeding back the $\bar{Q}$ output as shown in figure 8.9, and by using the clock line as the T-line.

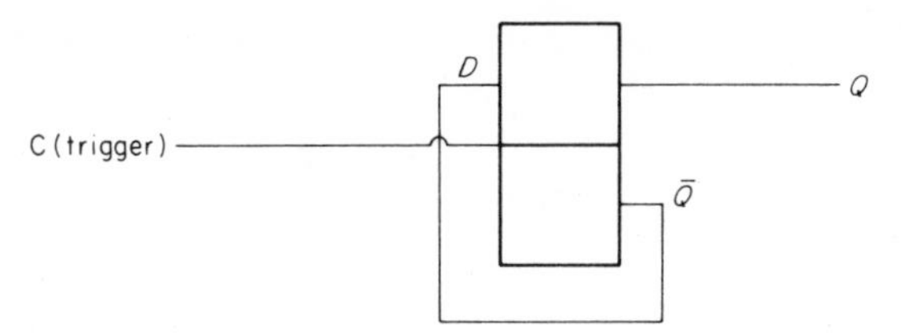

Fig. 8.9 A T flip–flop constructed from a D flip–flop

8.9 Edge-triggered flip–flops

In the master–slave flip–flops described above, the master flip–flop remains active (that is, it is connected to the input line) during the time that the clock signal is at the '1' level, and the data is finally transferred to the output when the clock signal falls to '0'.

In the range of flip–flops described as edge-triggered flip–flops, data is transferred to the output *on the incidence* of either the leading edge (positive edge-triggered) or the trailing edge (negative edge-triggered) of the clock pulse. Let us consider the operation of a positive edge-triggered flip–flop.

The data input (J, K, or D signals) must be applied to the input terminals for a minimum period of time known as the **set-up time** (typically 10 ns) prior to the clock pulse reaching its threshold value (typically 1.5 V in a TTL device). After the clock pulse has reached this value, the input information must remain for a period of time known as the **hold time,** which is typically 0–5 ns. After this point the input information is 'locked-out', and has no further effect until the clock pulse falls to zero again.

8.10 Dynamic memories

Dynamic memories depend for their operation on the ability of the parasitic capacitance of the gate of a MOST to retain an electrical charge. The simplest, and most popular form of MOS dynamic memory is the three transistor circuit of figure 8.10.

Transistor TR1 is used as the active device which stores the information, the parasitic gate capacitance C being the storage element. A logic '1' is 'written' into the memory by charging capacitor C by applying a negative voltage to the gate of TR2 via the **write select** line, and then by charging C to a negative voltage by energising the **data** line. After a period of time the charge on C decays a little, and the information is **refreshed** by writing the data in once more The information is, typically, refreshed every 2 ms or so. If we wish to discharge C, the data line is held at zero potential whilst the write select line is activated.

To **read** the data stored in the memory, TR3 is turned ON by the application of a negative voltage to the **read select** line, the state of the cell being detected by sensing the current flowing in the **data out** line. The **access time** to the information stored is, typically, 150 ns.

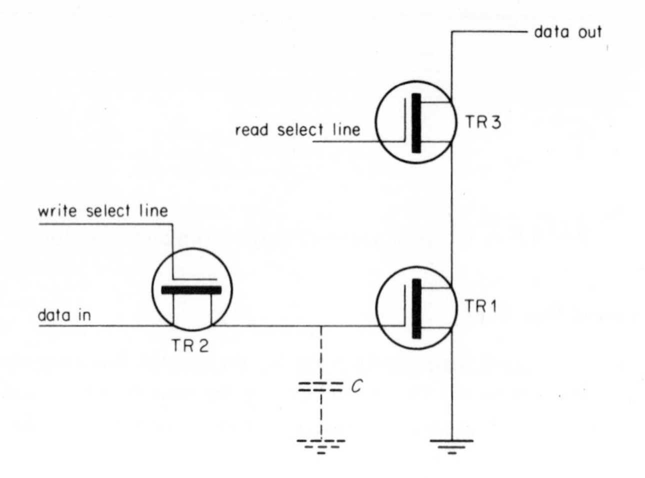

Fig. 8.10 A basic form of dynamic memory

8.11 RAM's, ROM's, and CAM's

The range of jargon used in conjunction with logic and computing is continually changing, but certain terms have been adopted as standard terminology and include RAM's, ROM's, and CAM's.

RAM

In logic jargon, a RAM is a **random access memory** and is a bank of memory cells in which an individual memory can be located by means of an **address** within the bank or matrix. Suppose there are nine flip–flops arranged in the matrix in figure 8.11. Each flip–flop has an address within the memory, and the address of the upper left-hand memory element is given as row 1, column 1, which we shorten to 1, 1. The flip–flop at address 1, 2 is located in row 1 of column 2. By energising row select wire 2 and column select wire 3, we 'address' the flip–flop FF 2,3 at address 2, 3.

In this way we can obtain access to each of the memory cells in a random fashion. Once we have obtained access to the memory, we can either write data into it or extract data from it.

ROM

A ROM is a **read-only memory** which permanently holds data, and which cannot be altered. A ROM could be used, for example, to contain a program of information which tests all the operations of an electrical typewriter.

The data contained in the ROM is often specified by the user, and is inserted into the ROM either at the manufacturing stage or, in the case of electrically programmable ROM's (that is PROM's), before installation into the equipment.

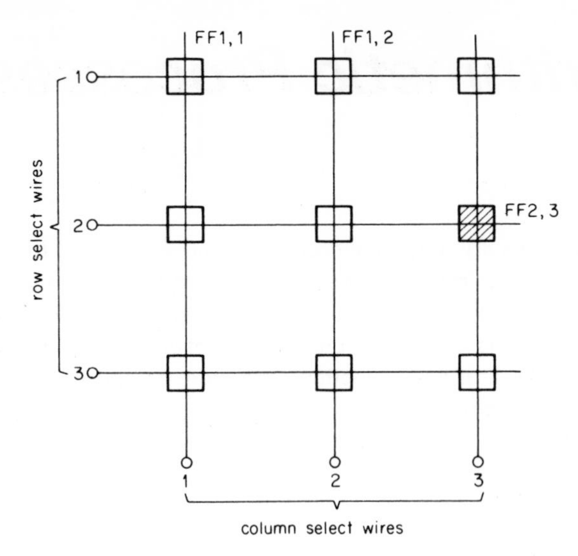

Fig. 8.11 The principle of a random-access memory

CAM

A memory which can be located on the basis of the contents it holds rather than an address within a memory matrix is known as a CAM, or **content addressable memory.** For example, when a car is registered it is necessary to define it in terms of the make, colour, year of registration, etc. Should we wish to determine the number of cars of a certain colour purchased in a given region of the country, this information can readily be obtained by comparing the *contents* of certain parts of each memory with reference data. To obtain the required data by scanning the complete memory would take a much longer time.

9 Arithmetic Processes

Electronic calculators are used in almost every walk of life, and the basic functions carried out by these machines are addition, subtraction, multiplication, and division. More complex machines carry out involved calculations such as taking square roots, generating logarithms, etc. In this chapter we concern ourselves with the principles of the four basic functions, the more complex operations being the concern of books on computers.

9.1 The binary system

Just as we express a decimal number as a series of multiples of powers of ten, it is also possible to express the same number as a series of multiples of powers of 2. For example, the decimal number 14 is represented in decimal form as

$$(1 \times 10^1) + (4 \times 10^0)$$

The same number expressed in binary form is

$$(1 \times 2^3) + (1 \times 2^2) + (1 \times 2^1) + (0 \times 2^0)$$

or, expressed in *binary-coded-decimal* form (see also section 9.10) the decimal number 14 is 8 + 4 + 2 + 0 = 14.

The total number of digits used as the basis of the numbering system is known as the **radix** of the system. The decimal (or denary) system uses the ten digits 0, 1, 2 . . . 7, 8, and 9, and has a radix of ten; the binary system uses the digits 0 and 1, and has a radix of two. The relationship between the two systems for the first sixteen values (zero to decimal 15) is given in table 9.1, the binary sequence shown being known as the **natural binary code.**

Each term in the pure binary code is given an equivalent decimal value or **weight,** which commences to the immediate left of the **binary point** (which is the binary equivalent of the decimal point) with unity value, and doubles its decimal value for each succeeding digit, that is, commencing at the binary point the weights increase in the sequence 1, 2, 4, 8, 16, 32, etc. A feature of the binary system is that each digit can assume only one of two possible values, corresponding to the '0' and '1' logic levels. Each binary digit is referred to as a **bit,** and a four-bit code group of the type in table 9.1 can deal with up to 2^4 = 16 different values.

Table 9.1

	decimal		binary			
decimal 'weight'	10^1 (10)	10^0 (1)	2^3 (8)	2^2 (4)	2^1 (2)	2^0 (1)
	0	0	0	0	0	0
	0	1	0	0	0	1
	0	2	0	0	1	0
	0	3	0	0	1	1
	0	4	0	1	0	0
	0	5	0	1	0	1
	0	6	0	1	1	0
	0	7	0	1	1	1
	0	8	1	0	0	0
	0	9	1	0	0	1
	1	0	1	0	1	0
	1	1	1	0	1	1
	1	2	1	1	0	0
	1	3	1	1	0	1
	1	4	1	1	1	0
	1	5	1	1	1	1

9.2 Fractional numbers

As with decimal numbers, binary numbers which are less than unity in value are represented by a series of multiples of powers of 2, the powers to which the radix is raised having a negative sign, as follows.

binary value	decimal value
$1 \times 2^{-1} = 0.1$	0.5
$1 \times 2^{-2} = 0.01$	0.25
$1 \times 2^{-3} = 0.001$	0.125
$1 \times 2^{-4} = 0.0001$	0.0625

Thus, the decimal number 6.625 is represented in binary form as 110.101, that is, 4 + 2 + 0 + 0.5 + 0 + 0.125.

9.3 Binary addition

Whatever the radix of the numbering system, the same mathematical processes are involved in the addition of two numbers. That is, if the sum of two numbers is less than the radix, we simply write down the sum. For example, if we add the decimal numbers 4 and 5 together, we write down the sum as 9. If the sum is greater than the radix, then we record the amount by which the sum is greater than the radix, and carry a '1' forward to the next higher column of the addition. It is important to note that when we add *two* decimal values together, the 'carry' is *either* zero *or* it is unity, and is never any other value. Hence, if we add the decimal numbers 9 and 8 together, we record a 7 and carry 1 to the next higher column of the addition. In the above case, the carry digit is known as the **carry-out,** C_O. This is carried forward to the next addition, when it is known as the **carry-in,** C_I.

Thus, the addition process can be regarded as consisting of two steps. In the first part we add the two numbers, which are known as the **addend** and the **augend,** and generate a *sum* and a *carry.* We then add to the sum the carry-in generated by the previous calculation. The two parts are described as 'half-additions', the net result being a full-addition. Electronic adding circuits use two half-adders (see section 9.4) to form a full adder. The complete addition process for binary numbers is illustrated in the following.

If we wish to add together in binary the two numbers which are equivalent to decimal 11 and 14, we proceed as follows:

decimal	binary	comment
11	1011	addend
14	1110	augend
	0101	first partial sum
	1010	first half-addition carry
	10001	second partial sum
	0100	second half-addition carry
25	11001	sum

We shall now consider the process step-by-step. In the 2^0 column we must first add $1 + 0 = 1$, giving a carry of zero. For convenience, this carry-out (0) is shifted one place to the left so that it will be in the correct position to be 'carried-in' to the 2^1 addition. In the 2^1 column we have the addition $1 + 1$ in binary. However, we cannot write down the number 2 as the sum, since the binary equivalent of 2 is 10 (see table 9.1). Thus $1 + 1 = 0$, carry 1. Once more the carry-out, this time a '1', is shifted to the left to become the carry-in to the 2^2 addition. This process is continued until we have formed the partial sum of the addend and the augend, together with the associated carry bits.

The first partial sum and the first half-addition carry are added together to give a second partial sum and a second carry. The addition of these two binary groups gives the correct sum of the two numbers.

9.4 Addition networks

In the above it was shown that the addition process could be treated in two parts, that is, as a combination of two half-adder stages. Let us consider the truth table for the first half-addition of the addend and augend, A and B, which is shown in table 9.2.

Table 9.2

inputs		sum	carry
A	B	S	C
0	0	0	0
0	1	1	0
1	0	1	0
1	1	0	1

Inspecting the relationship which exists between inputs A and B and the sum S, we see that

$$S = A \,.\, \bar{B} + \bar{A} \,.\, B$$

That is, a **sum** is generated when A is **not equivalent** to B. All we need, therefore, to generate the sum of two binary numbers is a **not-equivalent** gate.

Inspecting the relationship which exists between the inputs and the carry signal, C, we find that

$$C = A \,.\, B$$

Combining these relationships in the form of a logic circuit, the network in figure 9.1 generates both the sum and the carry associated with the addition. Alternatively, we could use the circuit developed in chapter 7 (figure 7.11(c)) which generates both the sum and the carry functions.

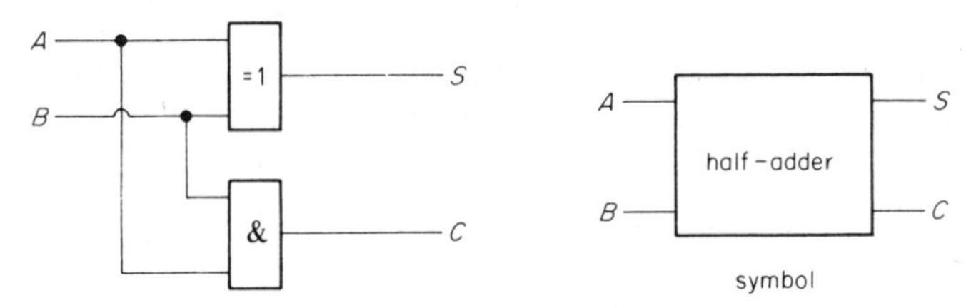

Fig. 9.1 A half-adder

To complete the addition process we need to be able to add the carry-in, C_I, to the sum, S, generated by the half-adder in figure 9.1. The complete addition will provide us with the **output sum**, S_O, and, quite possibly, a further carry bit, C_O. The generation of the two carry bits was illustrated in the example in section 9.3. A **full-adder** circuit which combines the outputs of two half-adders is shown in figure 9.2, the **output carry**, C_O, being obtained by OR-gating the carry outputs from the two half-adders.

A serial adder

The numbers A and B may be part of a sequence of binary digits which are presented to the adder from, say, the store of a computer. The essential elements of

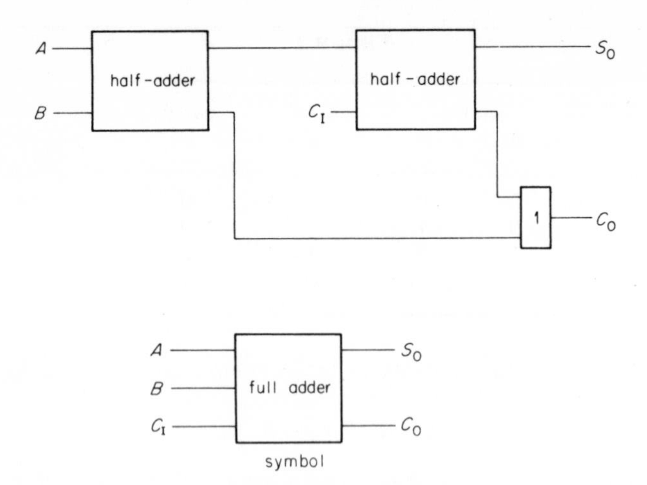

Fig. 9.2 A full-adder

the serial adder are shown in figure 9.3. In the case of a serial adder, the 2^0 digits of both numbers are presented to inputs A and B of the adder, the sum appearing as output S_O. The carry-out signal resulting from this stage of the addition is stored in the **carry store** FFC.

When the 2^1 bits of numbers A and B are presented to the input of the adder, they are added together with the previous value of the carry-out signal generated by the 2^0 addition. To delay the transfer of the C_O bit by the correct time interval in order that its arrival at the C_I input coincides with the 2^1 bits of A and B, the carry store is controlled by a clock signal. In figure 9.3 a D flip–flop is used as the carry store. More is said about the control of the flow of information in the following paragraph.

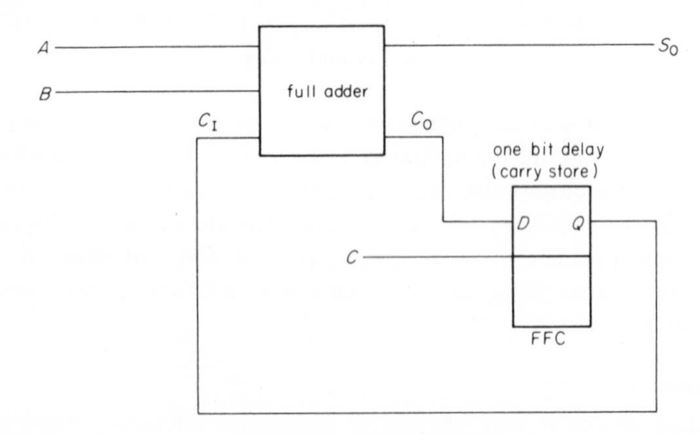

Fig. 9.3 A serial adder

In serial machines the numbers A and B and the sum S_O are stored in memory banks known as **shift registers** (see chapter 12), Each bit associated with numbers A and B is *shifted* into the adder under the control of a **clock pulse** or **shift pulse** C. By using a common shift pulse to shift data into inputs A and B as well as controlling the carry store, the correct delay is introduced to the carry bit so as to give the correct operation. If each number contains four bits, then four shift pulses are required to complete the addition.

A parallel adder

In some circuits the time taken to carry out mathematical operations must be kept to a minimum. Parallel addition is used in these cases, since, in parallel adders, all

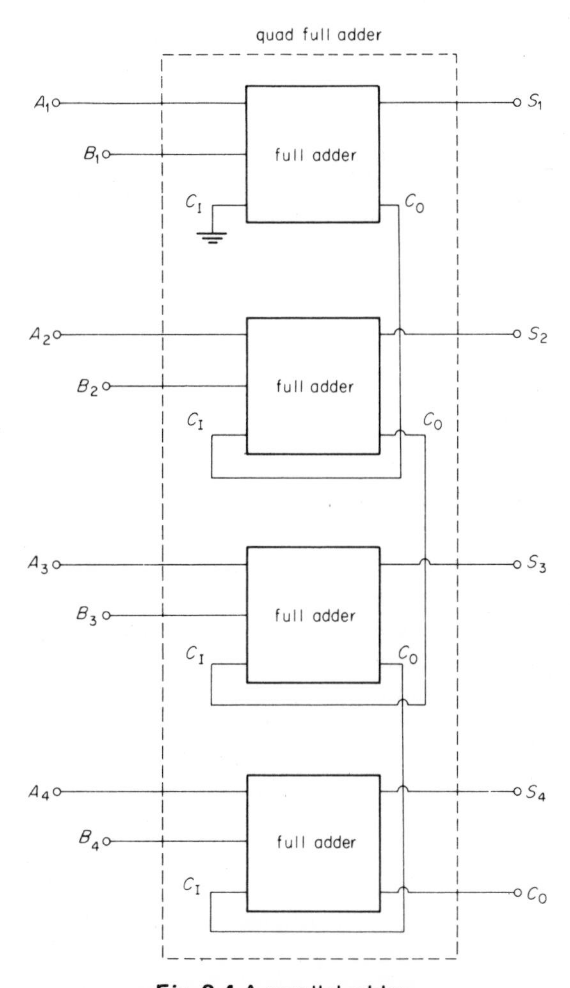

Fig. 9.4 A parallel adder

the binary digits are simultaneously presented to the adder and the addition is carried out in one operation.

A block diagram of a 4-bit parallel full-adder is shown in figure 9.4. A complete network of this kind in IC form is known as a **quad full adder,** and is contained in a 16-pin DIL pack. Inputs A_1, A_2, A_3, and A_4 correspond to the binary values of number *A*, A_4 being the **most significant bit** (m.s.b.). As can be seen from figures 9.3 and 9.4, parallel adders are more complex than serial adders, and are therefore more expensive to manufacture. In the network shown, as each value of *A* and *B* is added, the carry is immediately applied to the next higher group of digits. Thus, the time taken to add two numbers together is only the time required for the carry signal to ripple through four stages. Even so, computer designers have produced devices known as **carry look-ahead adders** which are even faster in operation. These adders include additional logic circuitry to reduce the number of stages through which the carry must be propagated.

9.5 Binary subtraction

The process of subtraction is carried out by converting the number to be subtracted (the **subtrahend**) into a negative number, and the *difference* is formed by *adding* the negative subtrahend to the **minuend**. This process is illustrated in section 9.6.

9.6 Negative numbers

A negative number is usually indicated by writing a 'minus' sign in front of the number. Unfortunately, electronic devices can only recognise the '0' and '1' signal levels, and cannot understand our minus sign. It is necessary, therefore, to develop techniques for the presentation of negative numbers in digital form.

The simplest method would be to use an additional bit in the code grouping, which we call the **sign bit,** which tells us whether the number is either positive or negative. We could, for example, allow the sign bit to assume the logical '0' value for positive numbers and allow it to be '1' for negative values. The remaining bits in the code group give the **magnitude** or **modulus** of the number. Such a number is said to be represented in **signed magnitude** (or **signed modulus**) form, several values being given below. In the examples given the sign bit is enclosed in brackets but, in a calculating machine, it would be identified simply by the fact that it is the m.s.b. in the number.

decimal	binary
−02.5	(1)0010.100
+ 10.25	(0)1010.010
−00.125	(1)0000.001

Readers will note the apparently unusual practice of including the zeros above the most significant digit in some of the numbers. These zeros are known as non-significant zeros, since they do not contribute to the value of the number. This has been done deliberately because, in calculators and computers, the memory stores the value of each and every bit, even if it is zero. In order to store a number as small as decimal 0.125 we need three bits to the right of the decimal point, and to store decimal 10 we need four bits to the right. In addition, a sign bit is also required. Thus to store numbers in the range ±10.125 we need a binary storage capacity of eight bits.

The signed modulus notation is a convenient method of storing numbers, but it does not allow us to directly subtract two numbers from one another. A number can be subtracted from another by *adding* its **complement,** which is its equivalent negative value. To illustrate the complement notation we shall study how it operates in the decimal system. Let us first of all determine the complement of the decimal number 1 by subtracting it from zero. Here we use a five-digit code, the m.s.b. being the sign digit with '0' representing a positive number and '9' representing a negative number.

```
(0)0000
(0)0001
        subtract
(9)9999 signed complement of decimal – 1
```

Hence we can differentiate between + 9999 and –1 by the fact that, in the case of –1, the sign bit is (9). Hence, if the sign bit is (9), then the number is stored in complement form.

Let us now look at the binary complement notation. Once again, to form the complement of the number, we subtract the positive value of the number from zero. This is illustrated below by forming the binary complement of 5 using a five-bit code, as follows:

```
zero        (0)0000
decimal 5   (0)0101
                    subtract
            (1)1011 signed complement form of – 5
```

The number so formed is known as the **true complement** or **two's complement** of the number. To verify that our result is correct, we shall subtract decimal 5 from decimal 9 in binary using two's complement addition.

```
decimal + 9                 (0)1001
decimal – 5                 (1)1011    signed 2's complement
                                       add
Overflow bit (lost) → 1(0)0100    = difference
```

The difference is (0)0100 or + 4 (decimal). Readers will note that in addition an overflow of '1' has occurred, but this is lost in the calculation since the storage capacity of our calculator is only five bits.

An alternative form of binary complement, known as the **one's complement** is also used. The 1's complement has a value which is less than the 2's complement by a factor of unity. The signed 1's complement of –5 is, therefore, (1)1010. Simple rules for the derivation of both the 1's and 2's complements are given below.

1's complement: Change all the 1's into 0's, and 0's into 1's.

2's complement: Commencing with the least significant bit (l.s.b.), leave all the digits up to and including the least significant '1' unchanged. All the more significant 0's are then changed into 1's, and 1's into 0's.
Alternatively, the 2's complement can be formed by adding '1' to the l.s.b. of the 1's complement.

Examples of the signed complement notation using a 7-bit data code are given below.

decimal value	binary modulus	signed 2's complement	signed 1's complement
– 6	(0)0110.000	(1)1010.000	(1)1001.111
– 12.5	(0)1100.100	(1)0011.100	(1)0011.011
– 15.875	(0)1111.111	(1)0000.001	(1)0000.000

9.7 Subtraction networks

As stated in section 9.6, we subtract a number B from number A by adding the 2's complement of B to A. A simple method of generating the complement was mentioned in section 9.6, which was to change all the 0's into 1's, and 1's into 0's via a NOT gate, and to add a '1' to the least significant bit. Having done this we can then use the full adder in figure 9.3 as a subtractor simply by feeding the 2's complement of B into the B input of figure 9.3. If the length of the number is n bits, we shall require an additional bit in the number length to account for the sign bit; that is, our serial subtractor system must have a storage capacity of $(n + 1)$ bits.

A feature of this form of subtractor is that we may be left with an *overflow* bit in the carry memory at the end of the calculation (see also example in section 9.6). It is therefore necessary to clear the carry memory before commencing each calculation in order to remove the surplus bit.

A 4-bit adder/subtractor using a quad full-adder is shown in figure 9.5. The add/subtract function is obtained by means of a control line which, on the application of a logic '1' to the line, causes the even-numbered AND gates to be opened (or **enabled**) and odd-numbered AND gates to be inhibited. This operation also causes a logic '1' to be applied to the C_I line of the adder, and causes the complements of the 'B' signals to appear at the B-inputs of the adder. In this way the 2's complement of number B is formed. Hence, when the control signal is '1', we subtract B from A. When the control signal is '0', the 'B' signals are applied directly to the B-inputs of the adder, so that A and B are now added together.

A complete serial subtractor is described in chapter 13.

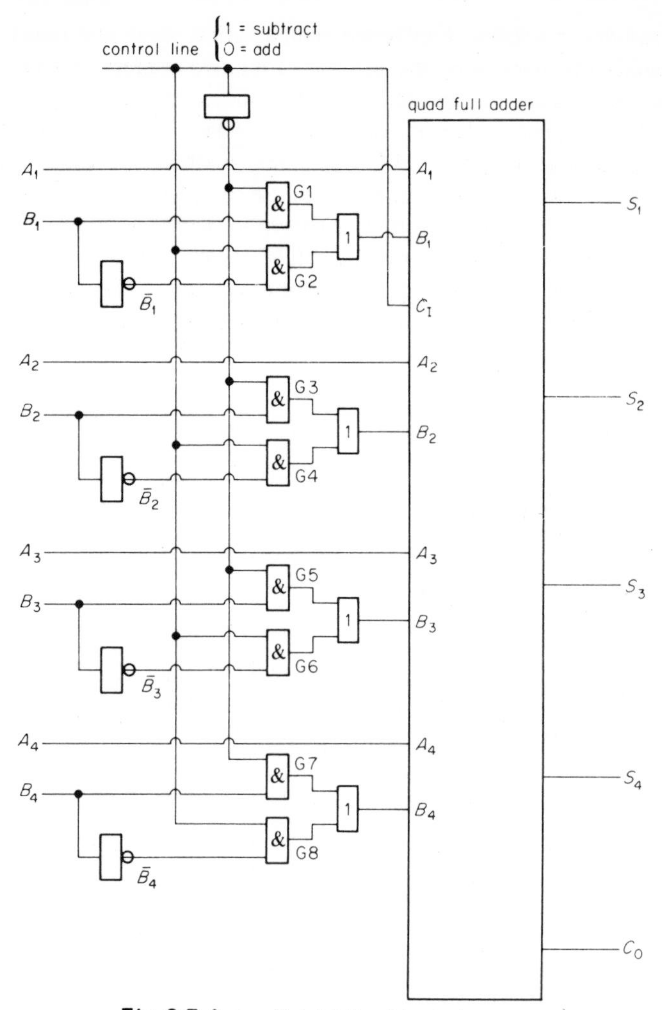

Fig. 9.5 A parallel 4-bit adder/subtracter

9.8 Multiplication networks

Multiplication can be performed by a process of adding and shifting, as illustrated in the following problem in which the binary equivalents of the decimal numbers 5 and 6 are multiplied together.

110	L (multiplicand) = 6
101	M (multiplier) = 5
110	partial product = L x 1
000	partial product = L x 0
110	partial product = L x 1
11110	product = sum of partial products

When multiplying numbers of different mathematical signs, the resulting sign bit can be generated by comparing the sign bits of the two numbers in a **not-equivalent** gate. If the sign bits are not equivalent to one another, the resulting sign bit will be a '1' (that is, the product has a negative sign). If they are equivalent, the output from the **not-equivalent** gate is '0'. Using this method, we must multiply the magnitudes of the two numbers and must not use complement notation.

A complete 4-bit parallel multiplier is shown in figure 9.6, in which the number $L_4L_3L_2L_1$ is multiplied by the number $M_4M_3M_2M_1$, where L_1 and M_1 are the least significant bits. Bits L_5 and M_5 are the sign bits. The decimal numbers associated with each output line is the 'weight' associated with a '1' signal at that output.

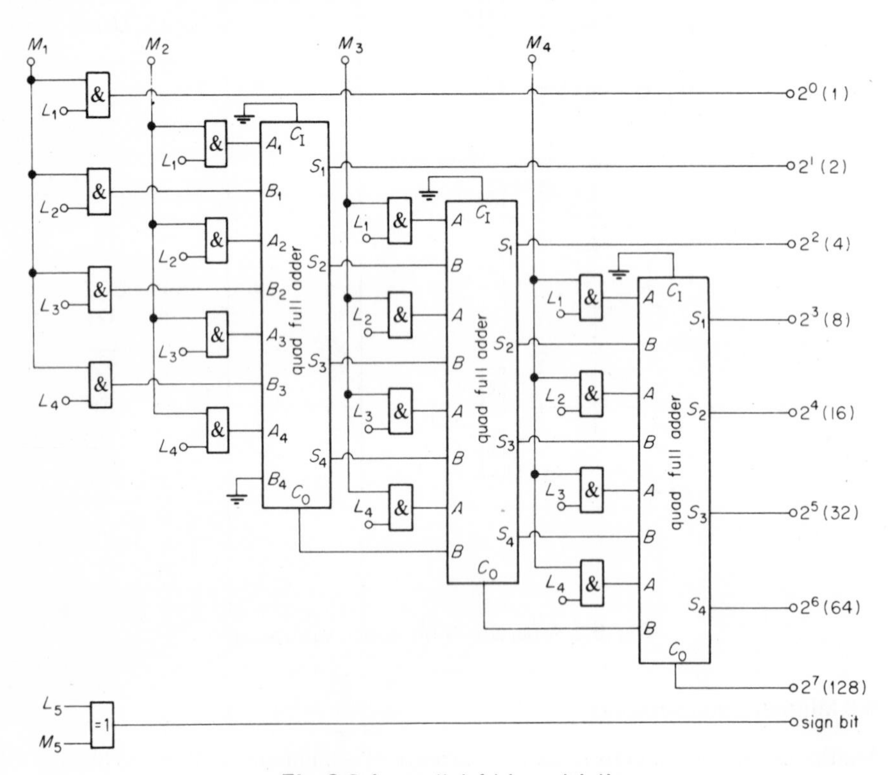

Fig. 9.6 A parallel 4-bit multiplier

The operation is as follows: Commencing at the left-hand side of the diagram, the first group of AND gates multiply each L-bit by M_1, which is the least significant M-bit, to form a partial product. The second tier of AND gates multiply number L by M_2, the resulting partial products being added in a quad full-adder. This process is repeated twice more, 'shifting' being introduced by the physical shift shown in the quad full-adders.

9.9 Binary division

As with decimal division, binary division can be performed by a process of subtracting and shifting. The design of a logic circuit for binary division is left as an interesting project for readers to attempt.

9.10 Binary–decimal codes

Where a man-to-digital system relationship exists, it is vital to establish codes which can be understood by both. The digital system can handle pure binary numbers with ease, but man experiences some difficulty in dealing with binary numbers. In an attempt to minimise the problems of communication involved, several codes known as **binary-coded-decimal (BCD) codes** have been developed. Using these codes, the presentation of decimal information in binary form is relatively simple. Three such codes are listed in table 9.3.

Table 9.3

decimal value	BCD codes								
	8421 BCD				excess-3 code	5421 BCD			
	(8)	(4)	(2)	(1)		(5)	(4)	(2)	(1)
0	0	0	0	0	0 0 1 1	0	0	0	0
1	0	0	0	1	0 1 0 0	0	0	0	1
2	0	0	1	0	0 1 0 1	0	0	1	0
3	0	0	1	1	0 1 1 0	0	0	1	1
4	0	1	0	0	0 1 1 1	0	1	0	0
5	0	1	0	1	1 0 0 0	1	0	0	0
6	0	1	1	0	1 0 0 1	1	0	0	1
7	0	1	1	1	1 0 1 0	1	0	1	0
8	1	0	0	0	1 0 1 1	1	0	1	1
9	1	0	0	1	1 1 0 0	1	1	0	0
unused combinations	1	0	1	0	1 1 0 1	0	1	0	1
	1	0	1	1	1 1 1 0	0	1	1	0
	1	1	0	0	1 1 1 1	0	1	1	1
	1	1	0	1	0 0 0 0	1	1	0	1
	1	1	1	0	0 0 0 1	1	1	1	0
	1	1	1	1	0 0 1 0	1	1	1	1

Each code uses ten of the sixteen possible combinations, the remaining six combinations being unused. The 8421 BCD code binary digits have weights of 8, 4, 2, and 1, respectively, and use the first ten groups of the natural binary code sequence listed in table 9.1. When reference is made to *the* BCD code, the 8421 BCD code is referred to.

The excess-3 code is generated by adding the binary equivalent of decimal 3 to each of the groups in the 8421 BCD code. An advantage of the excess-3 code over the 8421 BCD code is the ease with which some mathematical calculations can be carried out.

Yet another code, the 5421 BCD code, is shown in table 9.2. In this code the weight of the m.s.b. is five, so that the decimal number eight is made up of the group

$$(1 \times 5) + (0 \times 4) + (1 \times 2) + (1 \times 1)$$

Many other forms of BCD code exist, some having negative weights, the 642(−3) BCD code being an example. In the latter code the decimal number five is represented by the binary group 1011, and decimal seven by 1101.

9.11 Error detection

When a binary code group is being transmitted from one point to another within a system, it is possible that one or more bits may either be lost ('drop-outs') or picked up ('drop-ins'). When this occurs, the code group contains an error. Several codes have been developed which not only detect the errors but, in some cases, allow us to correct the errors. In the simplest form of error detection we introduce an additional bit to the code group, known as the **parity bit**, and which is **redundant** in terms of information transmission.

In an **odd parity check** system, the parity bit makes the total number of 1's in the code group an odd sum, and in the case of an **even parity check** it makes the sum an even number. Parity checks can be used with any type of code, and the 8421 BCD code together with the parity bit P for both odd and even parity are listed in table 9.4. The type of parity check selected depends on the application since each has its advantages; for example, a feature of an odd parity check is that no number is represented by a series of zeros.

Table 9.4

decimal value	odd parity 8421P	even parity 8421P
0	00001	00000
1	00010	00011
2	00100	00101
3	00111	00110
4	01000	01001
5	01011	01010
6	01101	01100
7	01110	01111
8	10000	10001
9	10011	10010

A parity bit generator for a code group which is presented in parallel mode, that is, all the bits are presented simultaneously, is shown in figure 9.7(a). The output from the NOT-EQUIVALENT gate G3 is the even parity bit associated with the four inputs. From table 9.4 we see that the odd and even parity bits are the logical complements of one another, so that the output from G4 in the figure is the odd parity bit associated with the four inputs.

A simple parity check circuit is shown in figure 9.7(b). Here, P_1 is the parity bit associated with the incoming code group, and P_2 is a parity check bit generated at the *receiving point* by a circuit of the type in figure 9.7(a). If the incoming parity bit is correct, then the output from the parity checker is zero.

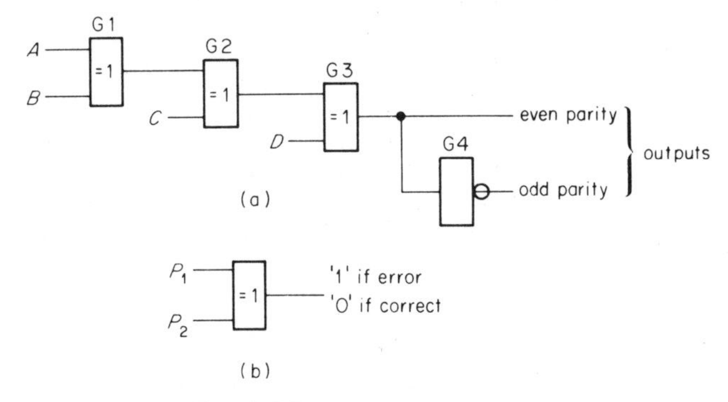

Fig. 9.7 Parity bit generators

A parity generator for a serial binary number could, quite simply, be a trigger flip–flop whose output is initially set to zero. The even parity bit is obtained from the Q-output of the flip–flop, and the odd parity bit from the $\overline{Q}$-output.

BCD illegal code checker

We saw in table 9.3 that out of the sixteen possible combinations of four bits, we needed only ten in the BCD code. The remainder are unused and are illegal code groups. Let us consider the design of a network which gives an indication of the existence of the illegal code groups in the 8421 BCD code.

From table 9.3 we note that the following groups should not exist: 1010, 1011, 1100, 1101, 1110, and 1111, the groups representing the sequence $ABCD$ where A is the m.s.b. Mapping these combinations on the Karnaugh map in figure 9.8(a), we see that the logical expression representing them is

$$\text{error} = A \,.\, B + A \,.\, C = A \,.\, (B + C)$$

A block diagram of a network which generates this function is shown in figure 9.8(b), the circuit giving a '1' output when an illegal code group is presented to its input. Readers will note that digit D in the BCD code is not required in this circuit.

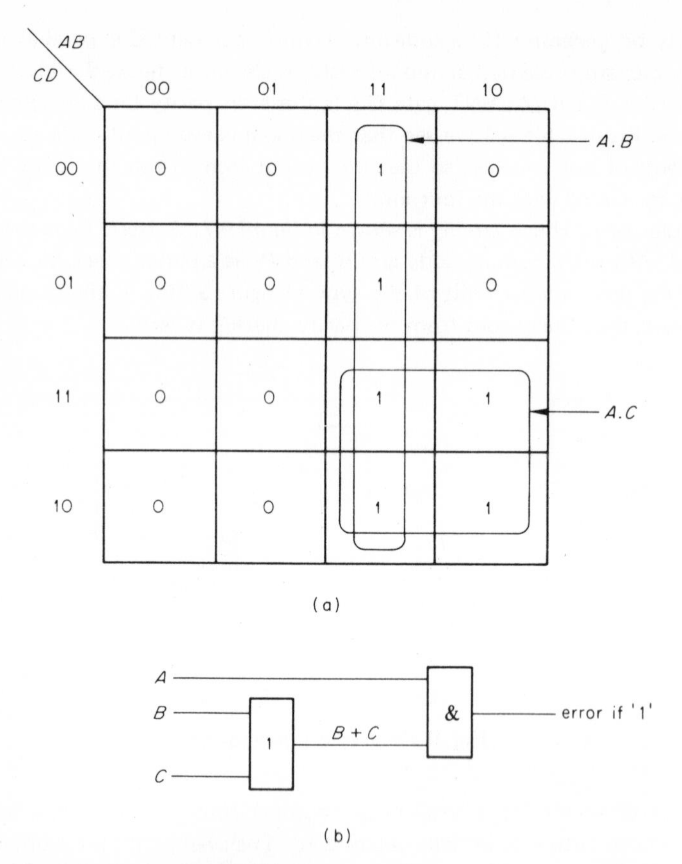

Fig. 9.8 An illegal code checking circuit for the 8421 BCD code

9.12 Binary comparators

Comparators are used in order to determine whether two binary variables A and B have the same value. If $A > B$, that is, $A = 1$, $B = 0$, then that fact can be determined by a logic gate which generates the function $A \cdot \overline{B}$, shown in figure 9.9(a). If $A < B$, that is, $A = 0$, $B = 1$, then a gate which develops the function $\overline{A} \cdot B$ (see figure 9.9(b)) generates a '1' output when this condition occurs. Referring to the **not-equivalent** gate in figure 7.11(a), we see that gate G1 in that circuit generates the $A > B$ function, and G2 generates the $A < B$ function. Thus,

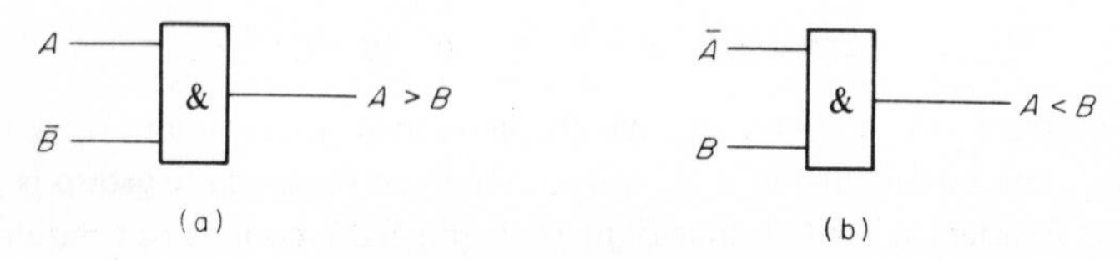

Fig. 9.9 Binary comparators

we can use some of the signals in the NOT-EQUIVALENT gate for the purposes of comparison of bits.

If we merely wish to determine whether two binary numbers are equivalent to one another, we could use an **equivalence** gate which develops the logical function $(A \cdot B + \bar{A} \cdot \bar{B})$.

There are many instances in digital systems in which we need to compare the values of two complete binary numbers in order to determine which is the greater of the two. One method of doing this is to compare the two numbers bit-by-bit, commencing with the most significant bits. Using a number of circuits of the type in figure 9.10(a), we can perform this operation. The $A > B$ and $A < B$ functions are generated inside the NOT-EQUIVALENT gate in the manner described above. For simplicity, the network in figure 9.10(a) is represented by the block diagram in figure 9.10(b), in which E is an *enabling* input signal, which allows A and B to be applied to the gates when $E = 1$.

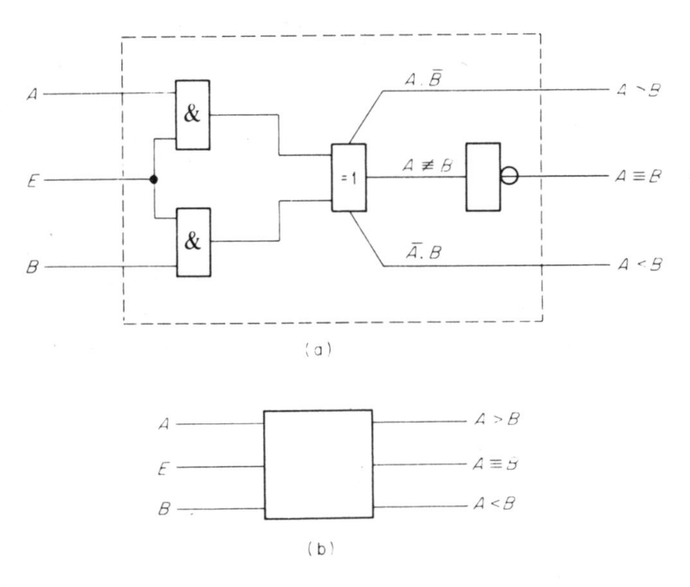

Fig. 9.10 A circuit providing both comparator and *equivalence* facilities

A complete network which compares two 3-bit numbers is shown in figure 9.11, in which A_3 and B_3 are the most significant bits of the numbers. The E-input of the m.s.b. comparator is connected to a logic '1' signal to allow the first stage of the circuit to operate at all times. If $A_3 > B_3$, then the output from the $A \equiv B$ line of the first stage is '0', and this inhibits the operation of the following stages. Gate G1 would then give a '1' output, indicating that $A > B$. If, on the other hand, $A_3 < B_3$, the following stages are once more inhibited and, in this case, G2 gives an output of '1'.

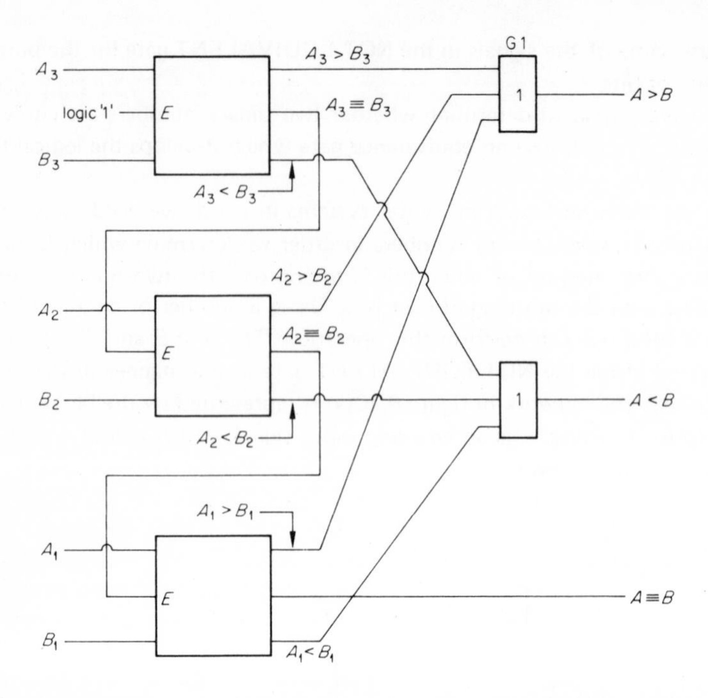

Fig. 9.11 A network which compares the values of two 3-bit numbers

When $A_3 \equiv B_3$, then the '1' output from the **equivalence** line of the first gate enables the operation of the second stage, which compares the values of bits A_2 and B_2 in the manner described above. If $A_2 \equiv B_2$, the circuit then proceeds to examine A_1 and B_1. Thus, the network allows us to determine whether A is either greater than, or less than, or equivalent to B.

10 Asynchronous Counters

10.1 A pure binary counter

When a group of flip–flops are connected together so that they store related information, they are known collectively as a **register.** Certain types of register can be used for the purpose of counting pulses, and are known as **counters.** An **asychronous counter** or **serial counter** is one in which the pulses are applied at one end of the counter, and the process of adding each pulse has to be completed before the 'carry bit' is propagated to the following stage. The following stage has then to add the carry bit to the number in that stage. That is, the carry bit appears to 'ripple' through the length of the counter until the count is complete. As a result, asynchronous counters are sometimes known as **ripple-through counters.**

A three-stage asynchronous counter using J–K flip–flops connected as T flip–flops (see also section 8.7) is illustrated in figure 10.1(a). All the stages of the counter are initially set to zero by the application of a signal to the reset line during the period of time when the input signal is zero (see section 8.6). As explained in section 8.6, the leading edge of the input pulse (that is, when the input signal changes from '0' to '1') has no effect on the state of FFC, so that all outputs remain at '0'. When the input pulse falls to zero, it causes the output of FFC to change to '1'. The change in signal at the output of FFC is applied to the 'clock' input of FFB but, as the change is a '0' to '1' change it has no effect on the operating state of FFB, that is, its output remains as '0'. The code sequence generated by the counter is listed in table 10.1, and the change described above corresponds to the change from the first row of the table to the second row.

At the end of the second pulse the output of FFC once more changes, this time from '1' to '0'. Since this corresponds to the trailing edge of the pulse applied to the clock input line of FFB, the output of FFB changes from '0' to '1'. This change in FFB output has no effect on the state of FFA, so that the outputs now are $A = 0$, $B = 1$, $C = 0$. Expanding the time scale associated with this transition as shown in the inset to figure 10.1(b), we see that it takes a finite time t_x for the transition to ripple through from FFC to FFB. This period of time is very small but, none the less, it is finite.

The count proceeds in the manner prescribed in table 9.1, the changes in the outputs occurring when the input to individual flip–flops changes from '1' to '0'. After seven pulses have been applied to the counter, all the outputs are at the '1'

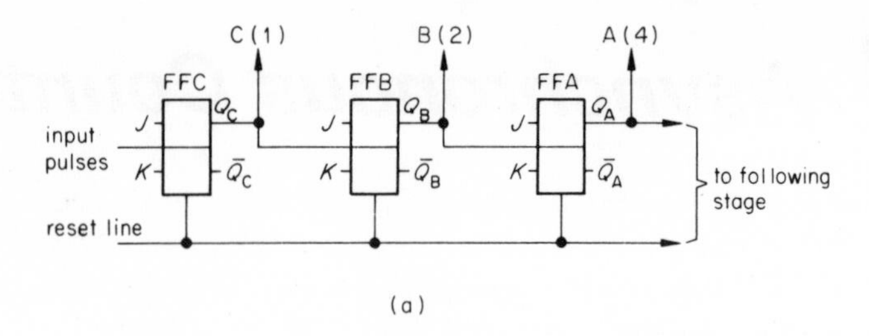

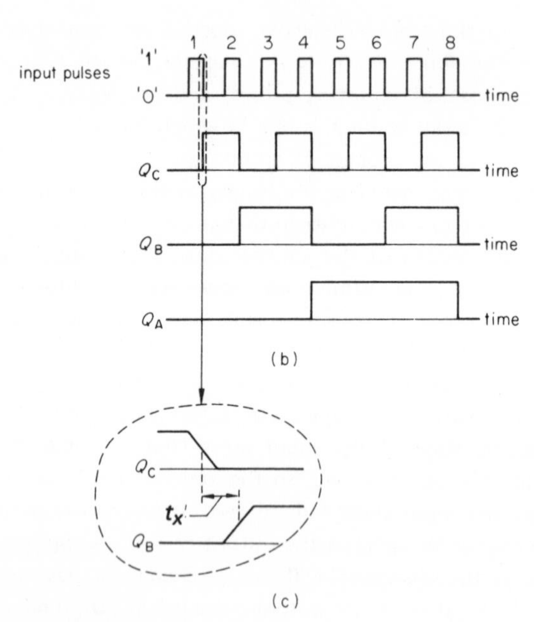

Fig. 10.1 A ripple-through pure binary counter

Table 10.1

pulse	A (4)	B (2)	C (1)	$\bar{A}$ (4)	$\bar{B}$ (2)	$\bar{C}$ (1)
Initial condition	0	0	0	1	1	1
1	0	0	1	1	1	0
2	0	1	0	1	0	1
3	0	1	1	1	0	0
4	1	0	0	0	1	1
5	1	0	1	0	1	0
6	1	1	0	0	0	1
7	1	1	1	0	0	0
8	0	0	0	1	1	1

(repeat: rows 8 → initial condition, for both column groups)

level, so that the eighth pulse causes them all to fall to zero. As shown in the inset to figure 10.1(b), the time taken for each change to propagate to the following stage is t_x, so that it takes $3t_x$ seconds before all the outputs from the counter have settled down to their steady-state values. As a result of this mode of operation, the maximum counting rate is restricted by the time delays in the counter, and for a counter with n stages the delay time before the final state of the counter can be 'read' after the application of a pulse is nt_x seconds. This time delay is very much reduced in the synchronous counters described in chapter 11.

10.2 A bidirectional pure binary counter

The type of counter described above is referred to as a **forward counter** or as an **'up' counter,** since it counts 'up' from zero. It is sometimes convenient to count from some predetermined value down to zero. A counter operating in this mode is known as a **reverse counter** or as a **'down' counter.**

The design philosophy of a 'down' counter can be deduced from table 10.1. Here we see that the **total** value stored in both the Q and $\bar{Q}$ outputs is constant and is equal to decimal 7 (this takes the weights of each bit into account). Thus, the initial value stored in the Q outputs is zero and that stored in the $\bar{Q}$ outputs is 7. After the first pulse, the Q outputs store 1 and the $\bar{Q}$ outputs store 6, and so on. Clearly it is possible to cause our counter to count 'down' either if we monitor the $\bar{Q}$ outputs rather than the Q outputs or if we use the $\bar{Q}$ output of an earlier stage to trigger the following stage. By the use of a suitable electronic gating system, we can design a **reversible counter** or **bidirectional counter** which can count either up or down.

A pure binary asynchronous reversible counter is shown in figure 10.2, in which the signal applied to line U controls whether the counter operates in the 'up' mode or the 'down' mode. If $U = 1$, the upper AND gates (that is, G1, G2, G3) are activated. Since one input of each of the upper AND gates is connected to a Q output line, then the counter counts 'up' when $U = 1$ since the changes in the Q outputs are gated forward. When $U = 0$, the lower AND gates (that is, G4, G5, G6) are activated. Since one input of each of these gates is connected to a $\bar{Q}$ output, then the counter counts 'down' when $U = 0$.

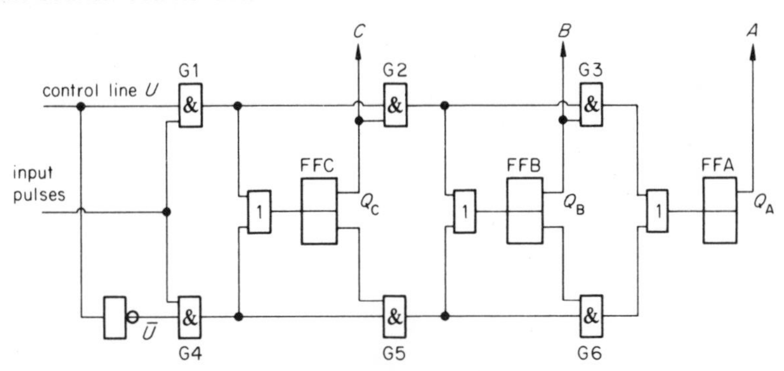

Fig. 10.2 A reversible pure binary counter

10.3 An 8421 BCD counter

A popular form of 8421 BCD 'up' counter using J–K flip-flops is shown in figure 10.3, and the code sequence generated is listed in table 10.2. Readers will note that FFB and FFD are connected to operate as conventional T flip-flops.

Table 10.2

pulse	ABCD
initial value	0000 ←
1	0001
2	0010
3	0011
4	0100
5	0101
6	0110 repeat
7	0111
8	1000
9	1001
10	0000 →

Initially, with all outputs reset to zero, the signal at the $\bar{A}$ terminal is '1'. This is fed back to the J and K inputs of FFC so that, initially, it operates in a T flip-flop mode. With the circuit so energised, the signal $B \,.\, C$ from G1 is zero, making FFA inoperative. Thus, FFD, FFC, and FFB operate as conventional trigger flip-flops, which then count 'up' in the normal pure binary sequence for the first seven pulses (see table 10.2). During this counting period (in fact after the sixth pulse), output $B \,.\, C$ becomes '1' and prepares FFA for operation. After the seventh pulse, the output from FFD once more becomes '1'. Since signal D is the clock source for FFA, it allows the '1' at the output of G1 to be entered into the 'J' master stage of FFA. At the end of the eighth pulse outputs B, C and D fall to zero. Since the

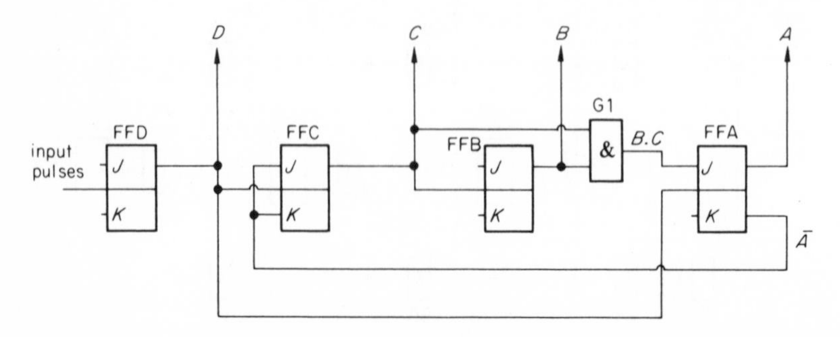

Fig. 10.3 A ripple-through 8421 BCD counter

change in the latter signal corresponds to the 'trailing' edge of FFA clock pulse, the '1' stored in the master stage of FFA is gated to output A. Thus after eight clock pulses $A = 1$, $B = 0$, $C = 0$, $D = 0$. At the same instant of time output $\bar{A}$ falls to zero, thereby inhibiting the operation of FFC and FFB. The ninth pulse causes output D to become '1' once more. The next pulse, the tenth pulse, results in D becoming zero again and this change is applied to the clock input of FFA. Since $B = C = 0$ at this time, the signal applied to the J-input of FFA is '0', so that the tenth input pulse causes a '0' ($=B \, . \, C$) to appear at output A. Thus, once more all the outputs are again zero, and the cycle is re-commenced by the eleventh pulse.

If FFA in figure 10.3 is replaced by a flip-flop with two J-inputs (see also figure 8.6), the necessity for gate G1 is removed, as the AND function is carried out inside the flip-flop.

11 Synchronous Counters

11.1 A reason for synchronous counters

In synchronous counters, the counting sequence is controlled by means of a clock pulse and all the changes of the outputs of **all** flip–flops occur in synchronism. This effectively eliminates the large propagation delay associated with ripple-through counters, as mentioned in chapter 10. Master–slave flip–flops are invariably used in synchronous counters to avoid the possibility of oscillation and instability when feedback connections are made in the completed counter. In this mode of operation, the appropriate input signals are simultaneously gated into the master stages of all the flip–flops in the counter. When the input pulse falls to the '0' level, the new values of the count are transmitted synchronously to the outputs of the flip–flops.

11.2 A synchronous pure binary counter

One form of synchronous pure binary counter is shown in figure 11.1, table 11.1 giving the sequence of events taking place in the counter, and from which we can deduce the design principles of the counter.

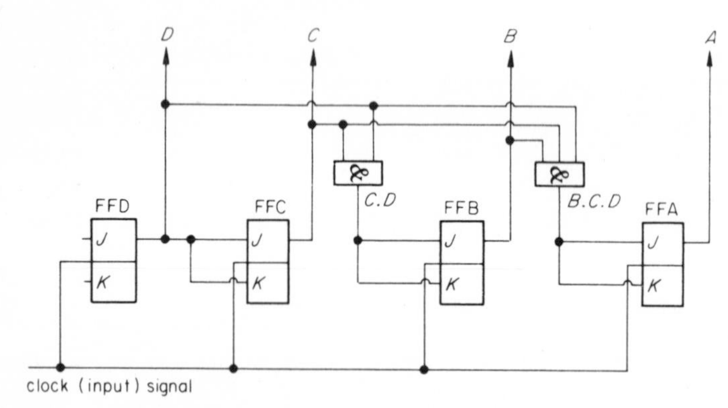

Fig. 11.1 A synchronous pure binary counter

We see from the table that the output of FFD must change after every input pulse. This calls for FFD to be connected to operate as a T flip–flop. The output of FFC is seen to change **following** the condition that $D = 1$ (that is, following

Table 11.1

pulse	A (8)	B (4)	C (2)	D (1)
initial condition	0	0	0	0
1	0	0	0	1
2	0	0	1	0
3	0	0	1	1
4	0	1	0	0
5	0	1	0	1
6	0	1	1	0
7	0	1	1	1
8	1	0	0	0
9	1	0	0	1
10	1	0	1	0
11	1	0	1	1
12	1	1	0	0
13	1	1	0	1
14	1	1	1	0
15	1	1	1	1
16	0	0	0	0

repeat

odd-numbered pulses). This change is brought about by driving both the J-input and the K-input of FFC by signal D. Also, the output of FFB must change state **following** the condition that the logical combination $C \,.\, D = 1$ is satisfied. This occurs after pulses 3, 7, 11 and 15 have been received. Similarly, the output of FFA must change **after** the condition $B \,.\, C \,.\, D = 1$ has been satisfied, that is, after pulses 7 and 15 have been received.

If J–K flip–flops with multiple J and K-inputs are used, then the AND gates associated with FFA and FFB are not required, as this function can be generated internally. For example, FFA together with its AND gate can be replaced by the IC package in figure 11.2, which uses a flip–flop with three J-lines and three K-lines.

11.3 A reversible synchronous pure binary counter

One form of reversible counter is shown in figure 11.3, the 'up' or 'down' mode being selected by the signal applied to line U which is the up/down control line. If $U = 1$, the counter operates in a count-up mode, and when $U = 0$ it counts down.

When $U = 1$, the '0' output from gate G1 inhibits the operation of gates G5, G6, and G7; the '1' signal on the U line permits G2, G3, and G4 to function. In this condition, the part of the circuit that is operational is generally similar to the counter in figure 11.1. Consequently the circuit operates in a count-up mode. When $U = 0$, gates G2 to G4 are inhibited and gates G5 to G7 are opened to the flow of information. In this operating state, the data stored in the $\bar{Q}$ outputs is transmitted forward, and the counter counts down.

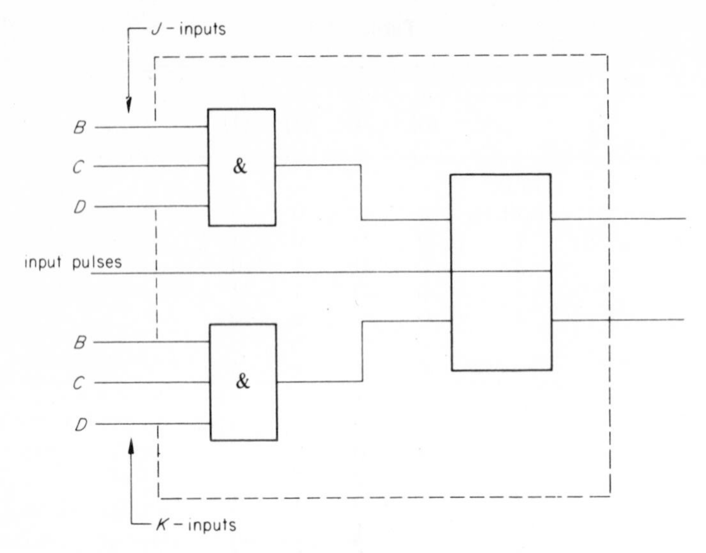

Fig. 11.2 The use of multiple-input J–K flip–flops

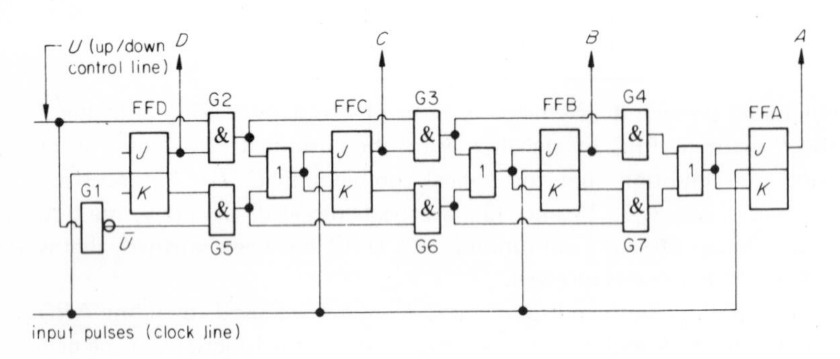

Fig. 11.3 A reversible synchronous pure binary counter

11.4 A synchronous 8421 BCD counter

The general arrangement of the synchronous counter in figure 11.4 is similar to the asynchronous BCD counter described earlier in section 10.3. The operation of the circuit is as follows.

Assuming that the outputs are all initially zero, the '1' signal at output $\bar{A}$ is fed back to G1 to allow it to pass signals. The circuit is then electrically equivalent to the synchronous counter in figure 11.1. The network counts up synchronously for the first eight pulses but, at the end of the eighth pulse, output A becomes '1' and $\bar{A}$ falls to '0'. This action inhibits the operation of G1 and prevents further signals from being propagated to FFC. The ninth pulse causes the output of FFD to change to '1', when the conditions in the counter are $A = D = 1, B = C = 0$. At the end of the tenth pulse the previous value of D, '1', is gated into the K-input of FFA

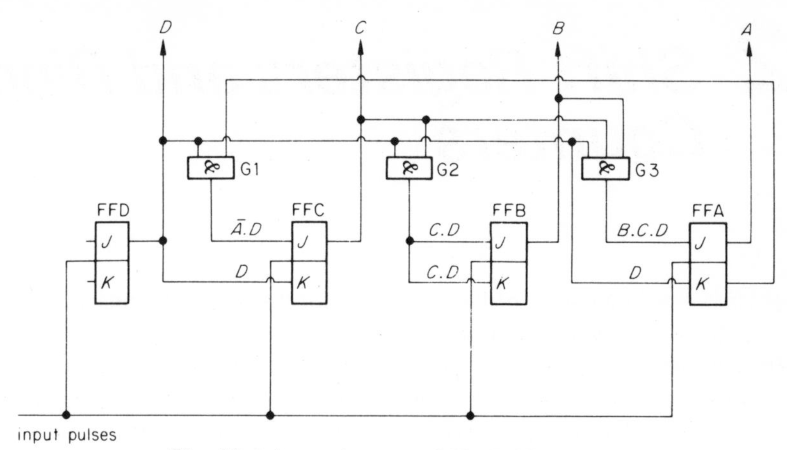

Fig. 11.4 A synchronous 8421 BCD counter

to cause it to fall to zero. At the same time, the tenth pulse also causes output D to fall to zero, and the counter is reset to its initial condition of $A = B = C = D = 0$.

11.5 Preset counters

In some applications, counters need to be **preset** to a particular value at the commencement of the count. This can be brought about by applying signals either to the set or clear lines of the flip-flop. One method of doing this is shown in figure 11.5.

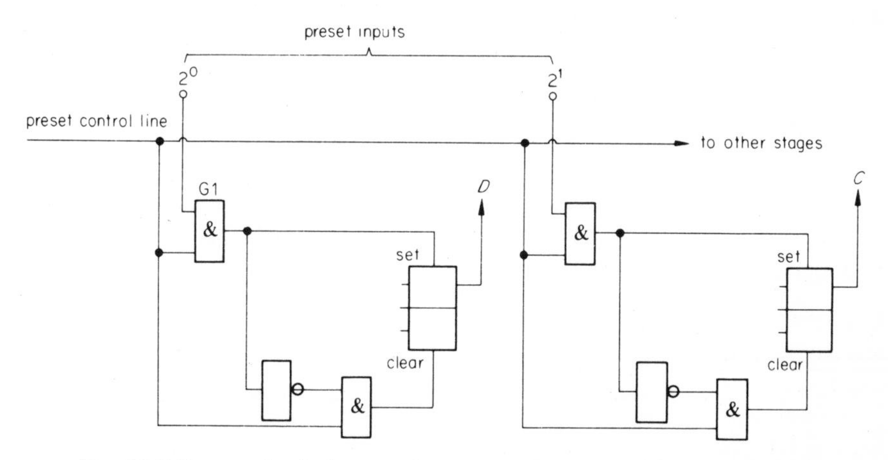

Fig. 11.5 One method of presetting counters to a pre-determined value

At a suitable stage in the counting cycle, the preset control line is activated by a '1' signal which allows the states of the preset inputs to be applied to the flip-flops. In the case shown, only the 2^0 and 2^1 preset controls are shown, but the circuit can readily be extended to deal with any number of input lines.

12 Shift Registers and Ring Counters

A register is simply an array of flip–flops used for the storage of binary data, and a **shift register** is one which is designed so that the data may be 'shifted' along the register in either direction, that is, either to the right or to the left. A **ring counter** is a shift register which is connected in the form of a continuous ring, the input to the 'beginning' of the ring being a logical function of the signals at one or more points in the register.

12.1 A serial-input, serial-output shift register

Two basic circuits for shift registers are shown in figure 12.1. In the circuits shown, the data to be stored in the register is shifted into the register in a serial fashion from the left-hand end, and is shifted out at the right-hand end Let us consider the operation of figure 12.1(a).

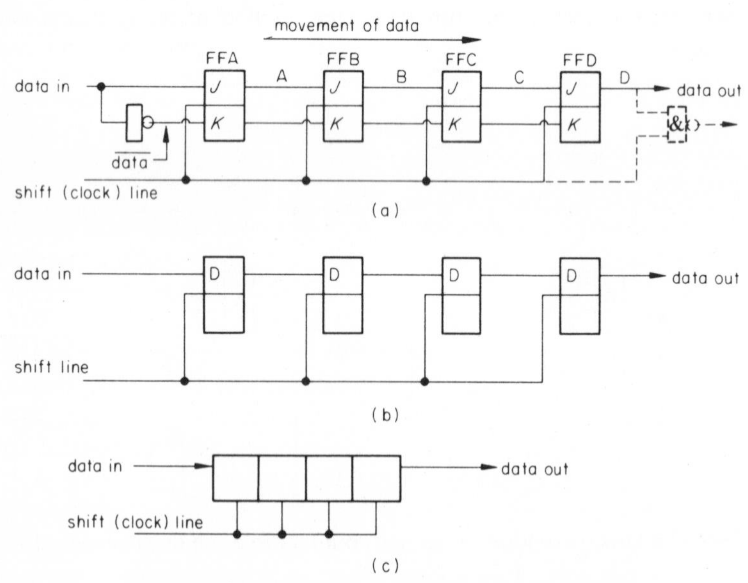

Fig. 12.1 Serial-input, serial-output shift registers

Assume that outputs *A, B, C,* and *D* are initially zero. The data presented at the input line is logically inverted by the NOT gate, so that complementary signals are presented to the *J* and *K* lines of the first flip–flop. Suppose now that a '1' signal is applied to the DATA IN line and that a pulse is applied to the clock (SHIFT) line of the register. At the end of the clock pulse the '1' is transferred to output *A*. Meanwhile, outputs *B, C,* and *D* remain at '0', since the input signals applied to those flip–flops were '0' during the period that the clock pulse was at the '1' level. The result of the above operation is shown in the first two rows of table 12.1. The '1' injected above is marked with an asterisk in table 12.1 to indicate its movement through the register.

Table 12.1

clock pulse	signal on data line	state of outputs *A*	*B*	*C*	*D*	
initial condition		0	0	0	0	
1	1*	1*	0	0	0	one complete shift cycle
2	0	0	1*	0	0	
3	0	0	0	1*	0	
4	1	1	0	0	1*	
5	1	1	1	0	0	
6	1	1	1	1	0	

Thus we have shifted a '1' from the data line into FFA. Also we have shifted the '0' formerly stored in FFA into FFB, as well as shifting the '0' in FFB into FFC, and the '0' in FFC into FFD. Hence *all* the information stored in the register has been transferred one step to the right.

If, now, the signal on the data line is reduced to zero and another shift pulse is applied, then this '0' is shifted into FFA at the end of the pulse. This change is brought about by shifting a '1' into the $\overline{A}$ output and at the same time the 1* is shifted into FFB.

Since the shift register has four stages, it can store a binary **word** of four bits, and requires four clock pulses to shift a new word into the register. Hence, after four clock pulses we have (i) serially shifted out of FFD the word 0000 and (ii) serially shifted in the word 1001. In the register in figure 12.1(a), we say that the information has been **shifted up** or **shifted to the right.**

We can continue to shift data into and out of the register, and table 12.1 is continued beyond one cycle to show the effect of shifting the first two 1's of the next word into the store. This has the effect of shifting the two right-hand bits stored at the end of the first cycle of the register.

In the form of circuit shown in figure 12.1(a), the output from FFD will be a continuous signal, either a '1' or a '0', for the whole of the clock cycle. If we need the output in the form of a pulse train, then it can be obtained by AND-gating the output of FFD with the clock pulse as shown in broken lines in figure 12.1(a). The interconnections between the flip–flops can be simplified by the use of D-type elements, as shown in figure 12.1(b). These flip–flops have internal NOT gates which generate the $\overline{\text{DATA}}$ information internally.

A symbol used to represent serial registers is shown in figure 12.1(c).

12.2 A reversible serial shift register

The data stored in the registers described above can only be shifted to the right. Some applications call for registers which can shift data either to the right or to the left at the command of a signal applied to a control line. The circuit in figure 12.2 is the basis of one type of reversible shift register. In this circuit, a '1' applied to the control line enables G1 to operate and it also inhibits G2. This permits data in the flip–flop to the *left* of FFB (FFA in figure 12.1) to be entered into FFB. When a '0' is applied to the right/left control line, G2 is opened and G1 is inhibited. This allows data stored in the flip–flop to the *right* of FFB (FFC in figure 12.1) to be shifted into FFB. Readers will note that the AND–OR combination in figure 12.2 can be replaced directly by three NAND gates (also see figure 7.6).

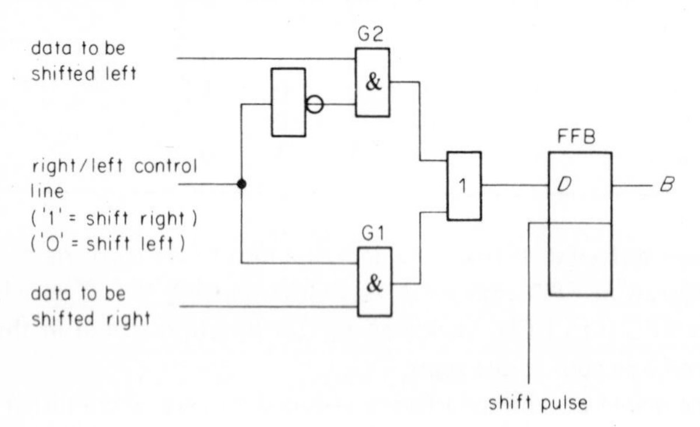

Fig. 12.2 One form of input arrangement to a stage of a reversible serial-input shift register

A three-stage reversible shift register using the circuit in figure 12.2 is shown in figure 12.3. When shifting right, data is fed in from the left-hand input into FFA, and is shifted serially into FFB, and then into FFC. When shifting left, data is fed serially into FFC from the right-hand input, and then is transferred into FFB and, finally, into FFA.

Once again, if the output is to be in the form of a pulse train, the appropriate outputs must be AND-gated with the clock pulse.

12.3 Parallel read-out

In some applications it is necessary to simultaneously observe the states of all the stages, that is, the output is required in parallel form. In the registers described so

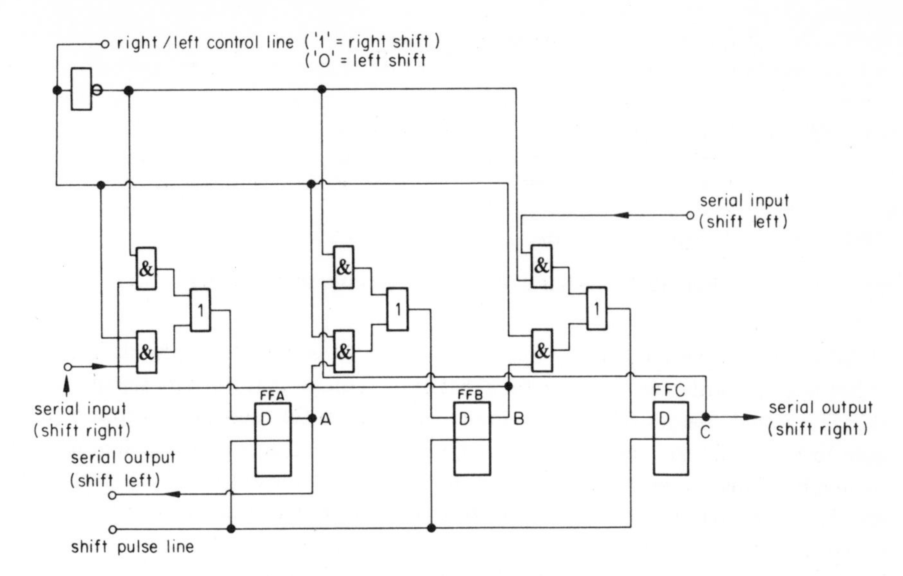

Fig. 12.3 A reversible shift register

far, this can be achieved if we employ a circuit in which the *Q* output of each flip-flop is available. In an IC pack this has the effect of limiting the maximum register length since, in the case of figure 12.1, we need a minimum of five pins on the IC pack, one each for the data in, data out, clock, power supply, and earth lines. In a 14-pin DIL pack there remain nine other connections for the additional input, output and control functions.

12.4 Parallel-input shift registers

As with binary counters, data can be preset into the register in a parallel mode. A logic network for carrying out parallel entry of data was described in section 11.5 and can be used in connection with shift registers.

12.5 Dynamic shift registers

A unique property of the MOSFET is the facility that it can be used as a dynamic memory element, as was described in section 8.10. When used in conjunction with other MOSFET's, simple and reliable shift registers can be constructed in monolithic IC form, the basis of one type being shown in figure 12.4.

The memory elements are the invertor stages G1, G2, G3, et cetera, together with the parasitic gate capacitors *C*1, *C*2, *C*3, et cetera. A complete shift register stage comprises two of these memory elements, together with two MOST switches which are turned ON and OFF by signals ϕ_1 and ϕ_2. The control lines ϕ_1 and ϕ_2 are energised by a **two-phase pulse supply**, the timing of ϕ_1 and ϕ_2 being such that the two switches associated with each memory are never closed simultaneously.

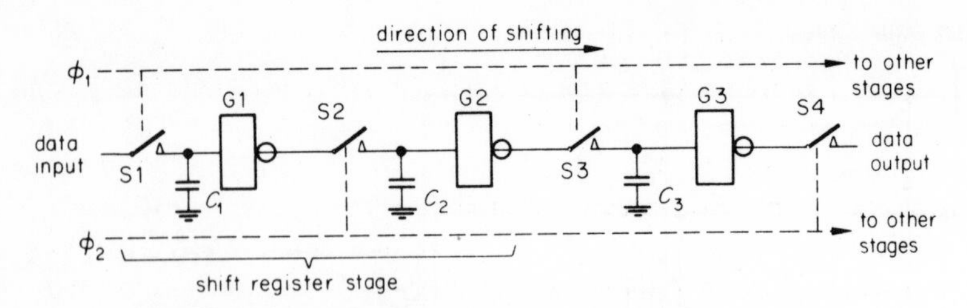

Fig. 12.4 The basis of a MOS dynamic shift register

When ϕ_1 is energised by a '1' signal, switch S1 is closed and the data on the input line is transferred to capacitor *C*1. Let us suppose that the data signal is logic '1'. This signal charges *C*1, and the inverting action of G1 causes the output of that gate to be '0'. At the end of the ϕ_1 pulse S1 opens, and a logic '1' signal is applied to line ϕ_2. This causes S2 to close, and the '0' signal at the output of G1 discharges *C*2. The inverting action of G2 causes the output of that gate to be logic '1'. Hence, in a complete cycle of events ϕ_1 and ϕ_2 become logic '1' in turn, and cause the input data to be transferred into the first stage of the register.

During the period of time that input data is transferred into G1, the data stored in G2 is transferred into G3. Clearly, signals ϕ_1 and ϕ_2 cause all the data stored in the register to be shifted to the right.

A diagram of one section of one form of dynamic shift register is shown in figure 12.5. In this circuit TR1 replaces S1 in figure 12.4, TR4 replaces S2, and TR7 replaces S3. Gate G1 in figure 12.4 is replaced in figure 12.5 by TR2 and TR3, and G2 is replaced by TR5 and TR6.

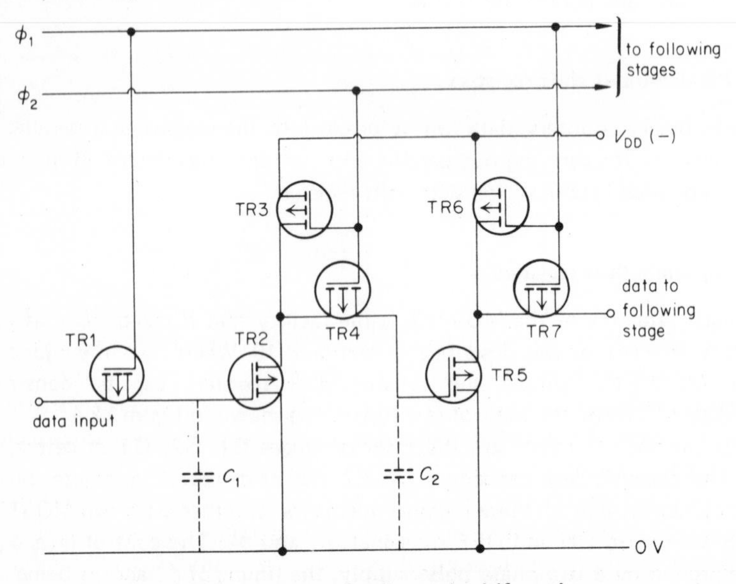

Fig. 12.5 One stage of a dynamic shift register

12.6 Ring counters

A ring counter is simply a shift register whose input is derived directly from its output, in the manner shown in figure 12.6.

Let us assume that stage A of figure 12.6(a) initially stores a '1', and that all other outputs are zero. The first shift pulse applied to the counter causes the '1' to move from stage A to stage B, and the 0's in stages B and C to move into stages C and D, respectively. Due to the feedback connection, the '0' in D is fed back into stage A. In this way the single '1' circulates continuously around the register in the manner shown in figure 12.6(b).

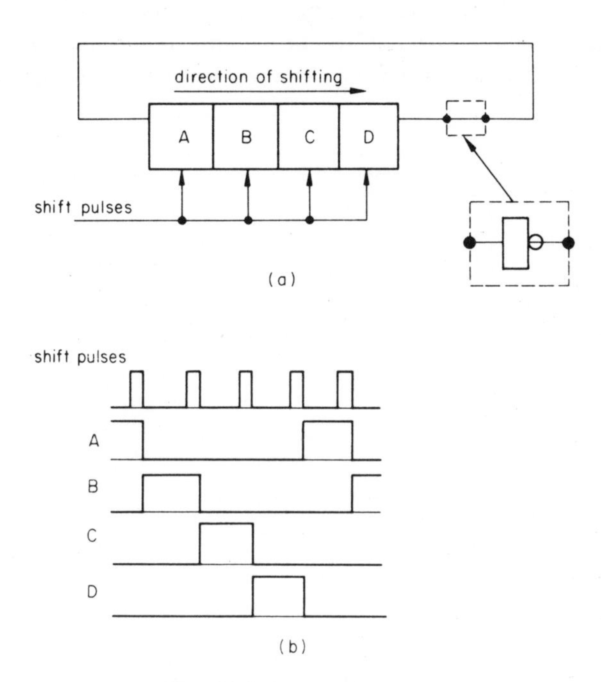

Fig. 12.6 A ring counter

A feature of the ring counter is that the value of the number stored is easily decoded, since when a '1' appears at output A it is equivalent to a count of zero, when it appears at B it is equivalent to a count of unity, when at C it is equivalent to a count of two, and when at D three pulses have been applied. This means that the **cycle length** of the code generated is four for a four-stage shift register. A ring counter which operates in a decimal code would need ten flip-flops.

As we have seen in the chapter on counters, it is possible to generate a cycle length of $2^4 = 16$ with four binary stages. Evidently, although the ring counter provides us with a simple method of 'reading' the number stored, it is uneconomic in its use of flip-flops.

Starting conditions in a ring counter

When the ring counter is first switched on, there is no guarantee that stage A will contain a '1' and that the other stages will contain zeros. A method sometimes adopted in circuits using this type of counter is to correct the situation at the end of the first complete cycle of operations by including an additional gate in the feedback network. The operation will be correct after the first cycle. The operation of such a circuit is outlined below.

We have seen that a '1' must be fed into stage A *following* the condition that the outputs of stages A, B, C, and D are *all* zero. If we use the feedback arrangement in figure 12.7(a) in which the input of stage A is energised by a NOR gate whose inputs are A, B, C, and D, then a '1' will be fed into A following the condition that $A = B = C = D = 0$. Alternatively we can use the feedback arrangement in figure 12.7(b).

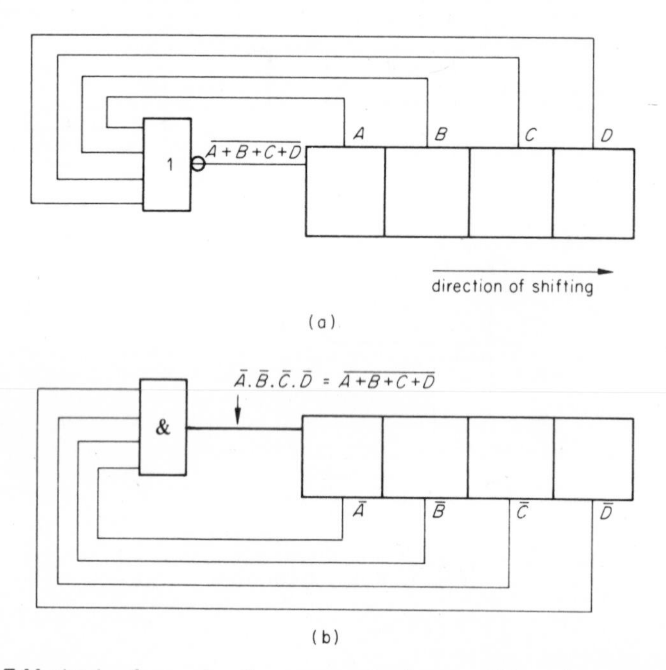

Fig. 12.7 Methods of ensuring the correct starting conditions in a ring counter

A twisted ring counter

By including the NOT gate shown in the inset to figure 12.6(a) in the feedback loop of the counter, the input to stage A is the logical function $\bar{D}$. This has the effect of increasing the cycle length to eight, that is, the code cycle length is double that of an otherwise equivalent ring counter. Offset against this is the fact that it is slightly more difficult to decode into decimal the binary value stored in the counter. The code sequence generated by the counter, known as a **twisted ring counter** due to the inversion in the feedback loop, commencing with the group 0000, is given in table 12.2.

Table 12.2

decimal value	*A*	*B*	*C*	*D*
0	0	0	0	0 ←
1	1	0	0	0
2	1	1	0	0
3	1	1	1	0
4	1	1	1	1
5	0	1	1	1
6	0	0	1	1
7	0	0	0	1
	0	0	0	0 →

In this case the logical complement of *D* is fed into A, so that *A* becomes '1' following the condition the *D* is '0', and becomes '0' following the condition that *D* is '1'.

Twisted ring counters are often known as **Johnson code counters**, and the code in table 12.2 is a 4-bit Johnson code. The correct logical conditions for a Johnson code sequence can be generated after one complete cycle by energising the input to stage C from a network which develops the logical function $B \,.\, (A + C)$, rather than by energising it directly by signal B. Readers may like to test their skill by verifying this fact.

12.7 Chain code generators

The binary sequence which is generated by a chain code generator has no apparent logical pattern and, in fact, appears to be a random sequence of binary numbers. The code is generated by a shift register whose input is derived either from a gate or from a network which develops a more or less complicated logical function of the outputs from the register.

A basic form of chain code generator is shown in figure 12.8, and comprises a 4-bit shift register whose input is supplied by the output from a NOT--EQUIVALENT gate. The input is fed in serially at the left-hand end, and the output may either be fed out in a serial or in a parallel mode. The code sequence generated by figure 12.8 is given in table 12.3, and we see from the table that a '1' is fed into stage A following the condition that $C \not\equiv D$.

Assuming that the cycle commences with $A = 1$, and all other outputs are equal to zero then, since $C \equiv D$, a '0' is fed into stage A when the first shift pulse is applied. Once more, after the first clock pulse, $C \equiv D$ so that a second '0' is fed into stage A by the second clock pulse. After this shift pulse condition $C \not\equiv D$ exists, and

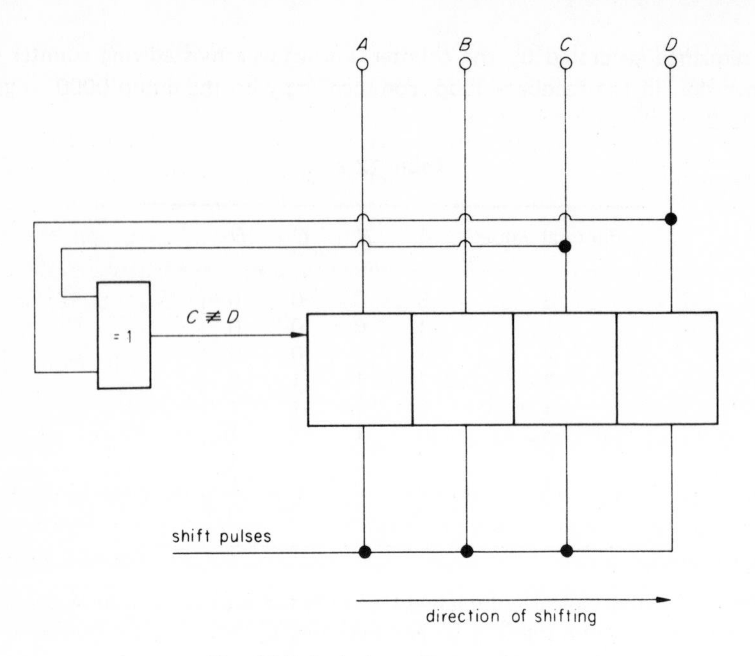

Fig. 12.8 A chain code generator

Table 12.3

shift pulse	*A*	*B*	*C*	*D*
initial condition	1	0	0	0 ←
1	0	1	0	0
2	0	0	1	0
3	1	0	0	1
4	1	1	0	0
5	0	1	1	0
6	1	0	1	1
7	0	1	0	1
8	1	0	1	0
9	1	1	0	1
10	1	1	1	0
11	1	1	1	1
12	0	1	1	1
13	0	0	1	1
14	0	0	0	1 →

a '1' is returned to the input of the register. This '1' appears at the output of stage A after the third clock pulse has been completed. The code sequence of the counter is completed after fifteen shift pulses have been applied.

This type of shift register counter utilises the flip–flops to greater effect than counters described hitherto since fifteen of the sixteen possible code combinations are used. The code group missing in table 12.3 is the combination 0000; if the outputs assume the 'all zero's' condition at the instant of switch-on, then there is no apparent movement of data since the zeros continuously circulate through the counter. In counters of this type it is necessary to include additional logic to prevent the 'all zero's' condition arising.

Had we used an equivalence gate in the feedback loop of figure 12.8, we should have generated a code sequence of length fifteen which includes the 'all 0's' condition, but which excludes the 'all 1's' state. Readers may like to verify this fact.

Since the pattern generated by the network is not a truly random pattern, we call it a **pseudo-random binary sequence** (PRBS). Applications of PRBS generators range from illumination control (for example, Christmas tree lighting) to electronic system testing.

The cycle length generated by chain code generators depends not only on the number of stages over which the feedback is applied, but also on the way in which the signal is derived. The maximum code length for N stages is $(2N-1)$, which is 7 for three stages, 15 for four stages, 31 for five stages, and so on.

13 Applications of Digital Electronics

13.1 Conversion of relay systems to electronic logic

A situation which arises in industry is the necessity for designing electronic logic systems to replace relay systems. Certain basic connections are repeatedly found in circuits used for the control of electrical machines, and in this section we shall show how the more important of these are converted into logic circuits.

Normally-open contacts in series

Where normally-open series connected contacts are used to complete a circuit, as shown in figure 13.1(a), they are replaced by an AND gate as shown in figure 13.1(b). This arrangement of contacts could occur in an installation in which switches A, B, and C must be closed before the circuit can be completed.

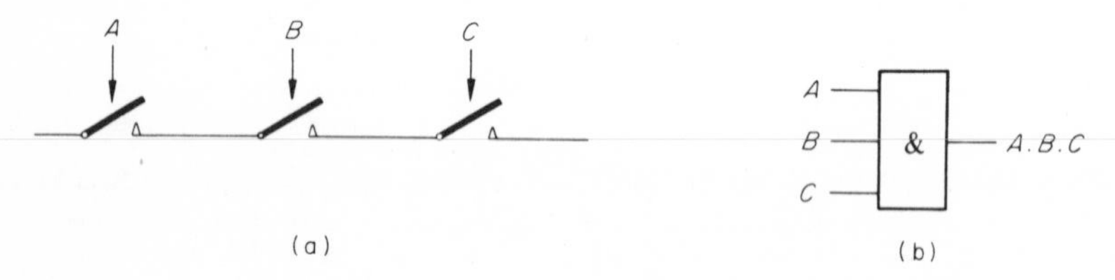

Fig. 13.1 Replacement of series-connected relay contacts by an AND gate

Normally-open contacts in parallel

Where we wish to start a motor by either of two switches, we connect the two in parallel as shown in figure 13.2(a). The electronic logic circuit which satisfies this arrangement is the 2-input OR gate of figure 13.2.

Normally-closed contacts in series

In some industrial applications it is necessary to include safety devices in the system, such as emergency stop push-buttons. These switches have normally-closed

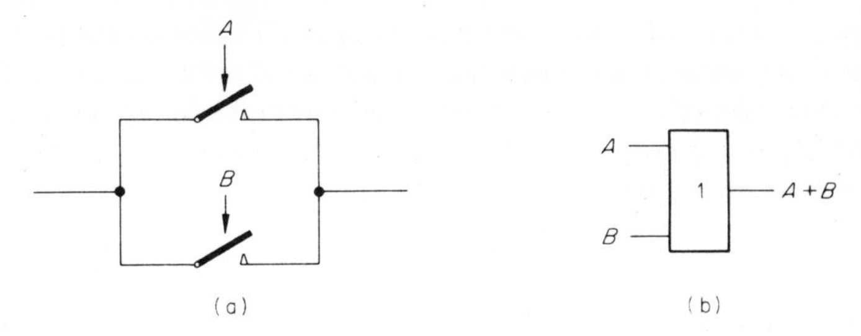

Fig. 13.2 Replacement of parallel-connected relay contacts by an OR gate

contacts which are opened when the switch is operated. This has the effect of cutting off the circuit current. The relay circuit is shown in figure 13.3(a), its logical equivalent being an AND gate with negated (NOT) inputs, as shown in figure 13.3(b).

Fig. 13.3 A method of replacing normally-closed relay contacts by a logiç gate

A complete circuit

Suppose that we are required to convert the relay circuit in figure 13 4(a) into an electronic logic network. Inspecting the circuit, we see that signals A, B, and C can

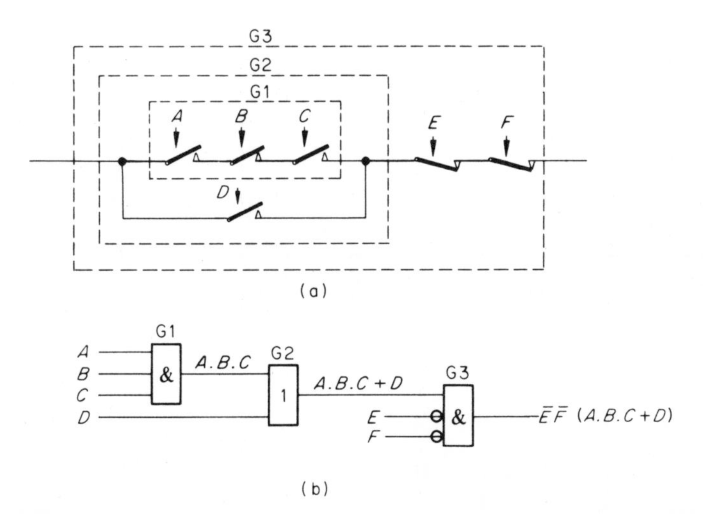

Fig. 13.4 The relay network in (a) can be replaced by the logic network (b)

be used to activate an AND gate, which we designate G1. The output from this gate is OR-gated with signal D, and this is brought about in the logic circuit by G2. Since the normally-closed contacts E and F are in series with the combined output of G2, we can AND-gate the output of G2 with the negated signals E and F. The complete logic block diagram is illustrated in figure 13.4(b).

13.2 Code converters

Quite frequently it is necessary to convert from a binary code into some other form of code; this occurs, for example, when we need a decimal read-out from an electronic counter which operates in binary. The interface between the two systems is known as a **code converter,** whose logical design can be deduced from a consideration of the truth tables of the code sequences.

13.3 8421 BCD code-to-decimal code converter

In this application we need to convert the 8421 BCD code into a decimal (that is, ten line) output for the purpose, say, of obtaining a display on a digital read-out device. The code sequence together with the decoding logic for the converter is given in table 13.1. Its design is discussed in detail below.

Table 13.1

decimal value (output)	8421 BCD code A (8)	B (4)	C (2)	D (1)	basic decoding logic	minimised decoding logic
zero	0	0	0	0	$\bar{A}.\bar{B}.\bar{C}.\bar{D}$	$\bar{A}.\bar{B}.\bar{C}.\bar{D}$
unity	0	0	0	1	$\bar{A}.\bar{B}.\bar{C}.D$	$\bar{A}.\bar{B}.\bar{C}.D$
2	0	0	1	0	$\bar{A}.\bar{B}.C.\bar{D}$	$\bar{B}.C.\bar{D}$
3	0	0	1	1	$\bar{A}.\bar{B}.C.D$	$\bar{B}.C.D$
4	0	1	0	0	$\bar{A}.B.\bar{C}.\bar{D}$	$B.\bar{C}.\bar{D}$
5	0	1	0	1	$\bar{A}.B.\bar{C}.D$	$B.\bar{C}.D$
6	0	1	1	0	$\bar{A}.B.C.\bar{D}$	$B.C.\bar{D}$
7	0	1	1	1	$\bar{A}.B.C.D$	$B.C.D$
8	1	0	0	0	$A.\bar{B}.\bar{C}.\bar{D}$	$A.\bar{D}$
9	1	0	0	1	$A.\bar{B}.\bar{C}.D$	$A.D$
	1	0	1	0		
	1	0	1	1		
	1	1	0	0	'can't happen' conditions	
	1	1	0	1		
	1	1	1	0		
	1	1	1	1		

The converter has four input lines *A, B, C,* and *D* which are simultaneously energised according to the sequence in table 13.1. At any particular state of the count, only one of the ten output lines is energised. Thus when $A = B = C = D = 0$, a logical '1' must appear on the 'zero' output line of the convertor (see figure 13.5); this signal is used to energise the 'zero' indicator of the read-out device. If the counter stores the binary equivalent of decimal seven, then $A = 0$ and $B = C = D = 1$, to give a logical '1' on output line 7; meanwhile the other decimal output lines are activated with logic '0' signals.

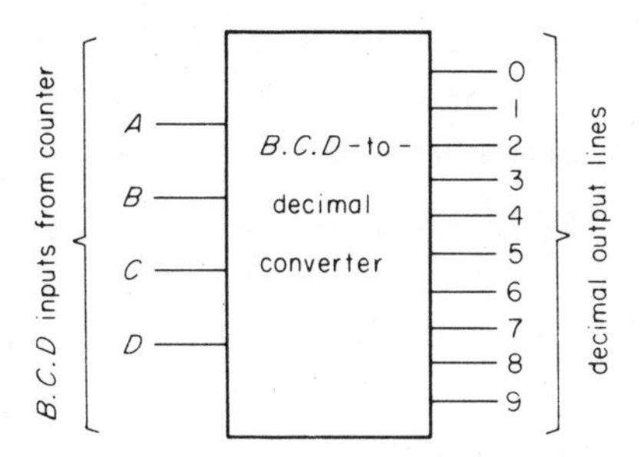

Fig. 13.5 An 8421 BCD-to-decimal code converter

From the above we see that the condition corresponding to a count of decimal zero is detected by a circuit which generates the function $\bar{A} . \bar{B} . \bar{C} . \bar{D}$. Similarly, a count of decimal 7 is detected by a gate which generates the function $\bar{A} . B . C . D$. The basic logical expressions for decoding BCD into decimal are derived in this way, and are listed in table 13.1, the complete decoder requiring ten 4-input AND gates. The logic for the code converter can be simplified if we map the basic code combinations in table 13.1 in the form of a Karnaugh map, as in figure 13.6(a).

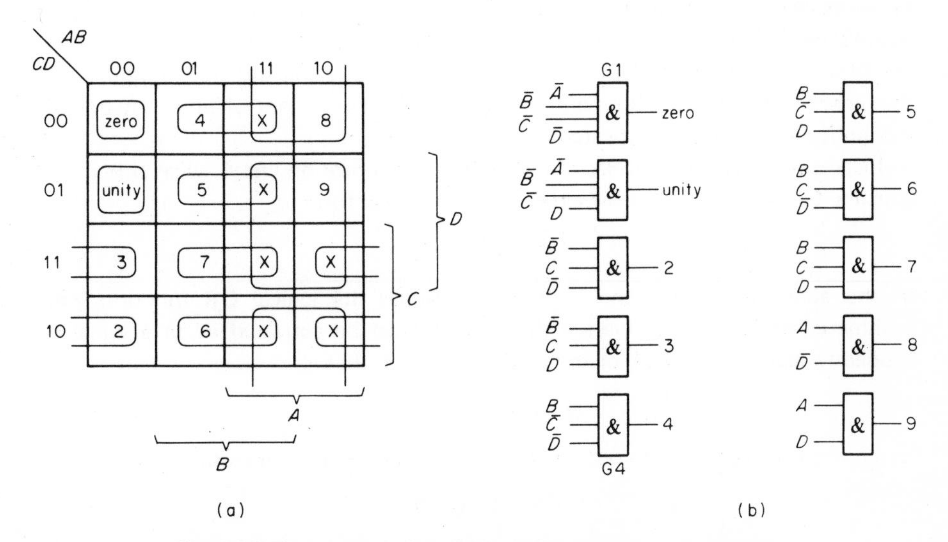

Fig. 13.6 The design of an 8421 BCD-to-decimal converter

In this map we write in each cell the decimal value corresponding to that cell. Thus in cell 0000 we write 'zero', in cell 0001 we write 'unity', in cell 0010 we write 2, and so on up to and including cell 1001 (decimal 9). The remainder of the cells are filled with X's, corresponding to the fact that they 'can't happen'. We then proceed to group each of the cells containing a number with the largest possible number of X's, being careful at the same time to exclude any other cell containing a number. The minimal expressions representing these groupings are also listed in table 13.1, and the AND network corresponding to the minimal logical statements is shown in figure 13.6(b).

BCD-to-decimal convertor using NOR gates

The AND gates in figure 13.6(b) can be replaced by NOR gates if each input signal is inverted. A NOR gate with inputs A, B, C, D can be used to replace G1, and one with inputs $\overline{B}, C, D$ can be used to replace G4, etc.

BCD-to-decimal convertor using NAND gates

We saw in section 7.3 that an AND gate can be constructed from two NAND gates, the final gate acting simply as an invertor. Alternatively, we may use a single NAND gate to replace an AND gate provided that we use the logic '0' level to identify the state of the count rather than use a logic '1' signal.

13.4 A Decimal-to-8421 BCD code converter

In computers and other digital calculators, we require a circuit which will accept a decimal input signal and convert it into the appropriate BCD code group. We may use table 13.1 once more, but in this case the ten decimal values are the inputs and the binary-coded-decimal lines are the outputs. At any instant of time only one of the input decimal lines is activated.

Inspecting table 13.1, we see that a signal must appear on line A when either line 8 or line 9 is activated. This is written down in logical form as

$$A = 8 + 9$$

In the above expression the 'pulse' sign implies the logical OR function, **not** arithmetic addition. Also we see that line B must be activated when a signal is applied to line 4 or to line 5 or to line 6 or to line 7, that is

$$B = 4 + 5 + 6 + 7$$

Extending this technique to outputs C and D, we obtain the expressions

$$C = 2 + 3 + 6 + 7$$

$$D = 1 + 3 + 5 + 7 + 9$$

Decimal-to-BCD convertor using NOR gates

A complete circuit for a decimal-to-BCD convertor using only OR gates is shown in figure 13.7. The gates in figure 13.7 can be replaced directly by the equivalent NOR network (see also chapter 7). Alternatively, each gate in figure 13.7 can be replaced directly by a *single* NOR gate with the proviso that an 'active' output is recognised as one which provides a '0' output, and not a '1' output.

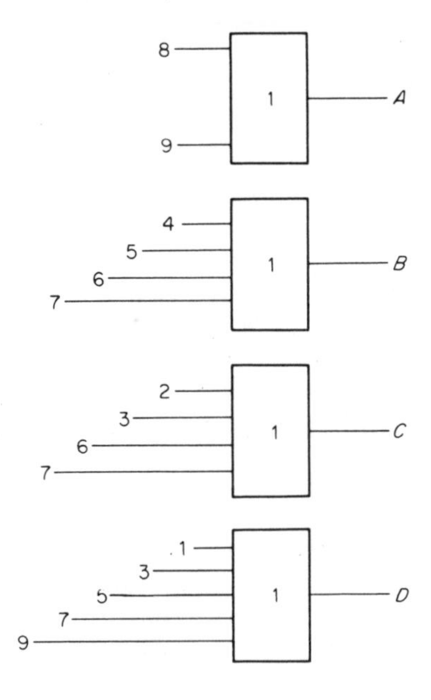

Fig. 13.7 The elements of a decimal-to-BCD converter

Decimal-to-BCD convertor using NAND gates

If we invert the individual input signals to the gates in figure 13.7 and we replace each OR gate with a NAND gate, then the resulting network is equivalent to figure 13.7.

13.5 A Johnson-code to seven-segment decoder

One of the most popular methods of displaying a decimal number is the 7-segment display. The basic format of this display is shown in figure 13.8(a), and consists of seven bars or segments, a to g inclusive, each of which may be either illuminated or extinguished, depending on the voltage applied to the segments. Displays corresponding to the ten decimal digits are shown in figure 13.8(b). Many types of device are used in this arrangement including light emitting diodes, liquid crystals, incandescent filaments, and gas discharge displays.

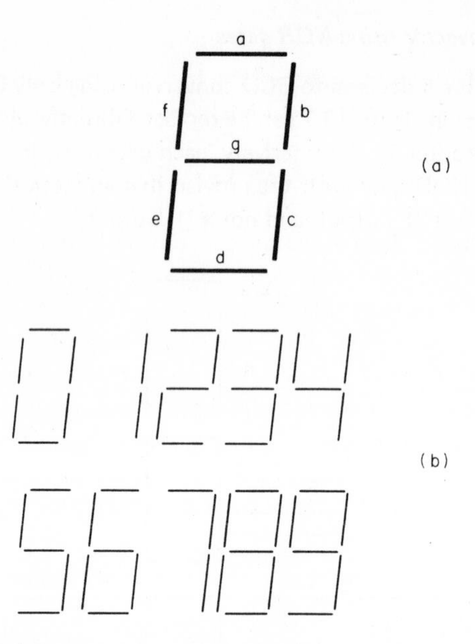

Fig. 13.8 Patterns generated by a seven-segment indicator

The code sequence necessary to generate the numeric indications in figure 13.8(b) is fairly complex. It is usual to carry out the pulse counting sequence in one of the more popular binary codes and then to convert that code into the 7-segment display code. In the following we consider the design of a code convertor capable of converting between a 5-bit Johnson code (see also section 12.6) and the 7-segment display code.

The Johnson code is generated by a twisted ring counter, and a five-stage register generates a code of length 2 x 5 = 10 code groups, that is, a decimal code. The Johnson code is widely used in digital instruments due to the simplicity of the counting circuit.

The basis of a 7-segment display stage is shown in figure 13.9. It contains a 5-bit Johnson code counter whose output is fed into a **data latch.** This is simply a group of five D-type flip–flops whose function it is to prevent the display from flickering

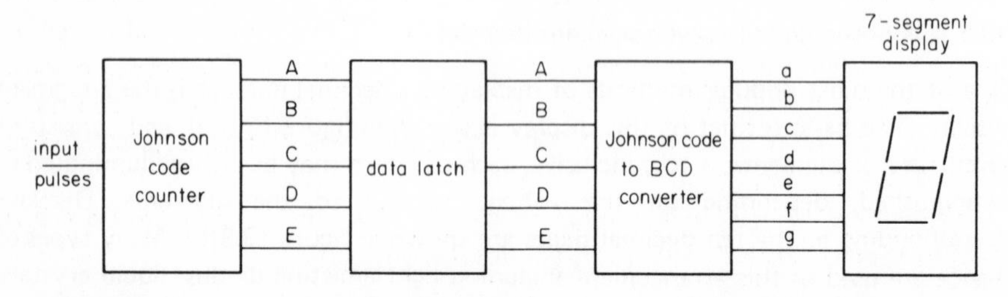

Fig. 13.9 A block diagram of a Johnson-code to seven-segment code converter

when the counter is operating. Where the flicker is not a nuisance, the data latch need not be used. At the completion of the counting cycle, the new value held in the counter is transferred into the data latch, and thence to the 7-segment display via the code convertor.

The code convertor is designed from table 13.2, which gives both the Johnson decimal code and the code required to display the decimal output. In the process outlined below, we firstly deduce the logical relationships necessary to convert the Johnson code into a decimal code and then, using those logical relationships, we design a decimal-to-seven segment decoder.

In order to detect the decimal 'zero' count, we need a 5-input AND gate which generates the function $\bar{A} . \bar{B} . \bar{C} . \bar{D} . \bar{E}$. Inspecting this relationship more carefully, and comparing it with the other combinations in the sequence we see that the minimal statement which *uniquely* defined the decimal 'zero' count is simply $\bar{A} . \bar{E}$. This process is repeated for each decimal state to give the minimal logical relationships for the Johnson-to-decimal code convertor.

Table 13.2

5-bit Johnson counter output					decimal value	Johnson-to-decimal logic	7-segment display code						
A	B	C	D	E			a	b	c	d	e	f	g
0	0	0	0	0	zero	$\bar{A} . \bar{E}$	1	1	1	1	1	1	0
1	0	0	0	0	unity	$A . \bar{B}$	0	1	1	0	0	0	0
1	1	0	0	0	2	$B . \bar{C}$	1	1	0	1	1	0	1
1	1	1	0	0	3	$C . \bar{D}$	1	1	1	1	0	0	1
1	1	1	1	0	4	$D . \bar{E}$	0	1	1	0	0	1	1
1	1	1	1	1	5	$A . E$	1	0	1	1	0	1	1
0	1	1	1	1	6	$\bar{A} . B$	1	0	1	1	1	1	1
0	0	1	1	1	7	$\bar{B} . C$	1	1	1	0	0	0	0
0	0	0	1	1	8	$\bar{C} . D$	1	1	1	1	1	1	1
0	0	0	0	1	9	$\bar{D} . E$	1	1	1	1	0	1	1

We now need to convert the decimal signals into a 7-segment code in order to derive the elements of the indicator. The 7-segment code sequence is obtained by inspecting the states of each segment in figure 13.8(b) for each of the decimal values. When zero is indicated, segment g is extinguished and all other segments are illuminated, so that we write a '0' in the g column of the 'zero' row of the code, and 1's in all the other rows. When the indicator displays 'unity', segments b and c are illuminated and all others are extinguished. Accordingly, we write 1's in the b and c columns of row 'unity', and 0's in all other rows. This process is repeated until the table has been completed.

Utilising the technique developed in section 13.4, we see that segment a is illuminated when the following condition is satisfied:

$$a = \text{'zero' or 2 or 3 or 5 or 6 or 7 or 8 or 9}$$

This expression requires an 8-input OR gate to provide the correct output for segment *a*. Alternatively, we note that segment *a* is *extinguished* at the counts of 'unity' and 4. Hence segment *a* must be illuminated when the logical condition NOT ('unity' OR 4) is satisfied. This condition is detected by a 2-input NOR gate, and is therefore used in preference to the OR-gate arrangement. Proceeding through the code in this way, and inserting the Johnson-to-decimal code conversion, we find that the overall convertor logic is

$$a = \overline{\text{unity} + 4} = \overline{A\,.\,\bar{B} + D\,.\,\bar{E}}$$

$$b = \overline{5 + 6} = \overline{A\,.\,E + \bar{A}\,.\,B}$$

$$c = \bar{2} = \overline{B\,.\bar{C}}$$

$$d = \overline{\text{unity} + 4 + 7} = \overline{A\,.\,\bar{B} + D\,.\,\bar{E} + \bar{B}\,.\,C}$$

$$e = \text{zero} + 2 + 6 + 8 = \bar{A}\,.\,\bar{E} + B\,.\,\bar{C} + \bar{A}\,.\,B + \bar{C}\,.\,D$$

$$f = \overline{\text{unity} + 2 + 3 + 7} = \overline{A\,.\,\bar{B} + B\,.\,\bar{C} + C\,.\,\bar{D} + \bar{B}\,.\,C}$$

$$g = \overline{\text{zero} + \text{unity} + 7} = \overline{\bar{A}\,.\,\bar{E} + A\,.\,\bar{B} + \bar{B}\,.\,C}$$

13.6 A time-division multiplexer

Time-division multiplexing (T.D.M.) is a technique used to convert data presented in parallel form, that is, data presented simultaneously on a number of lines, into sequential form so that it may be transmitted along a single wire. Examples of the use of this technique include the transmission of multiple telephone conversations along a single pair of wires, and also in data logging systems which scan a large number of input signals and print the data sequentially on a typewriter.

The basis of a T.D.M. system is shown in figure 13.10. Data input lines *A, B, C,* and *D* are connected to four fixed contacts on a switch. As the moving contact of the switch rotates, it sequentially scans the states of the data lines and transmits the data along the transmission line in the sequence *A, B, C, D.* In addition to the data, it is also necessary to transmit a synchronising signal in order that the receiving equipment can identify the commencement of the sequence.

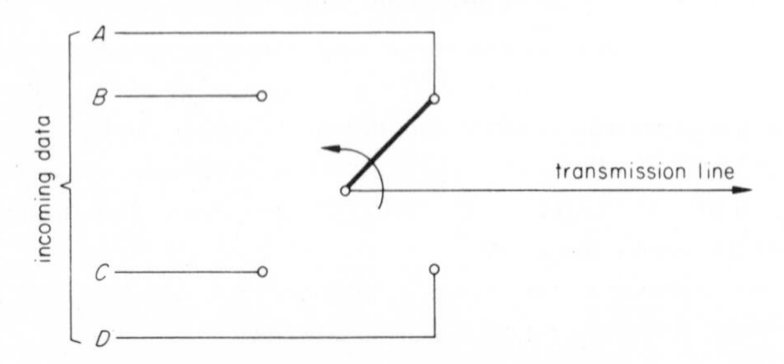

Fig. 13.10 The basis of time-division multiplexing

A simplified T.D.M. system using electronic logic is shown in figure 13.11. Here we use a ring counter to replace the rotating switch, the logic '1' which continuously circulates around it serving as the scanning contact. When the '1' appears in stage V of the counter it activates G1, so that the signal applied at input line *A* (which may either be a '1' or a '0') is transmitted along the DATA line. The next clock pulse shifts the '1' in the ring counter into stage W, causing the state of input *B* to be transmitted down the DATA line. Each time a clock pulse is received it causes the next input line in the sequence to be scanned. Synchronisation is achieved in this circuit by the use of a second line whose function it is to transmit a pulse when the '1' appears at stage Z of the counter.

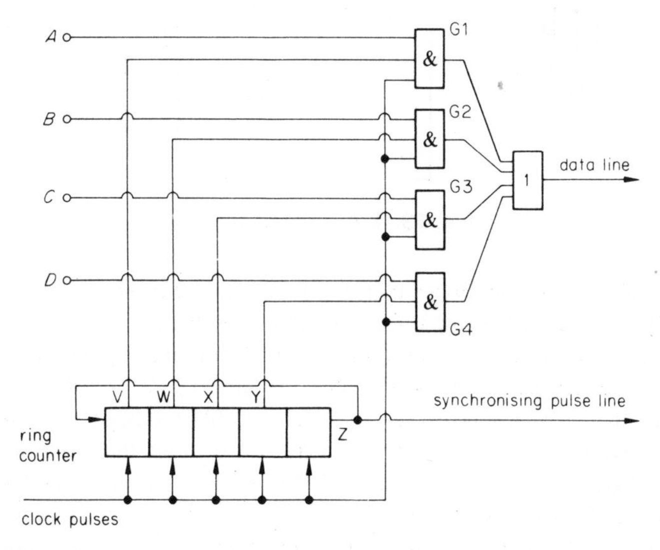

Fig. 13.11 A logic block diagram of a time-division multiplexing system

Readers will note that the clock pulse is also applied to the inputs of G1 to G4. This link is required only in the case where we need to introduce a 'guard' space between the data pulses. If the clock line were disconnected from these gates, and if $A = B = C = D = 1$, then a continuous voltage at the logic '1' level would appear on the DATA line.

13.7 A serial binary adder

The basis of serial addition was dealt with in chapter 9, and here we describe a complete circuit as may be used in a calculating machine.

Registers A and B in figure 13.12 are used to store the two binary words to be added together, and these are shifted serially into the full adder. At the same time, the sum formed by the addition is shifted into the *sum register* S, the carry for the following addition being stored in the carry store, the latter initially storing a '0'.

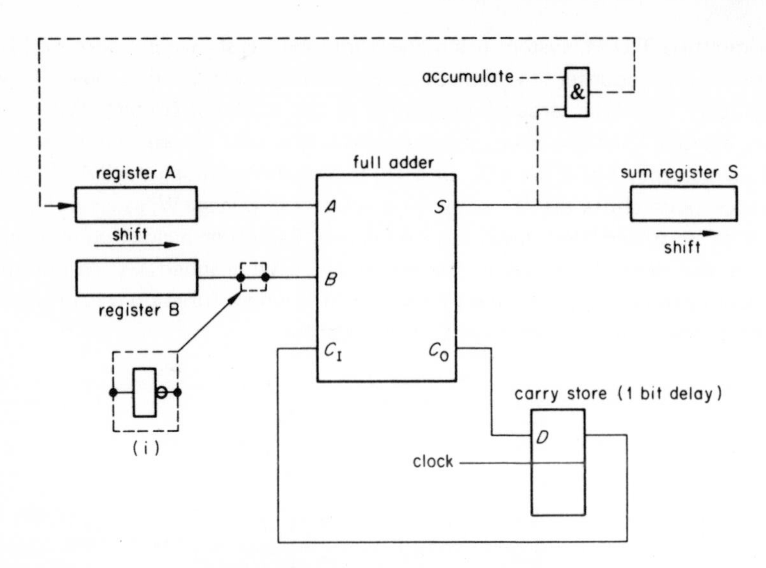

Fig. 13.12 A serial binary adder

The sum register must be longer than the binary number by one bit in order that any 'overflow' produced during the calculation is not lost.

The number of registers used can be reduced from three to two by including the recirculating circuit shown in broken line in figure 13.12. In this case, when the **accumulate** line is energised by a '1' signal, the SUM output from the adder is shifted into register A. Thus, register A carries out the dual functions of storing one of the numbers to be added, and also of storing the final sum. As the number in register A is shifted out of the right-hand end and into the full adder, so the sum is shifted into the left-hand end. When a '0' signal is applied to the accumulate control line and shift pulses are applied, a series of 0's are shifted into register A.

13.8 A serial binary subtractor

We saw in section 9.6 that subtraction can be carried out by the process of adding the complement of one of the numbers. If, in figure 13.12, we include the invertor shown in inset (i), the value accumulated in register S is the binary value of $(A - B)$. In this case the most significant bit stored in register S is the sign bit.

As with the adder, register A can be used for the dual functions of storing the minuend before the numbers are subtracted and the difference after the subtraction.

13.9 A pulse train generator

In some applications it is convenient to be able to generate a pulse train of N pulses, the value of N being controlled by a digital signal. The basis of one form of pulse train generator is shown in figure 13.13.

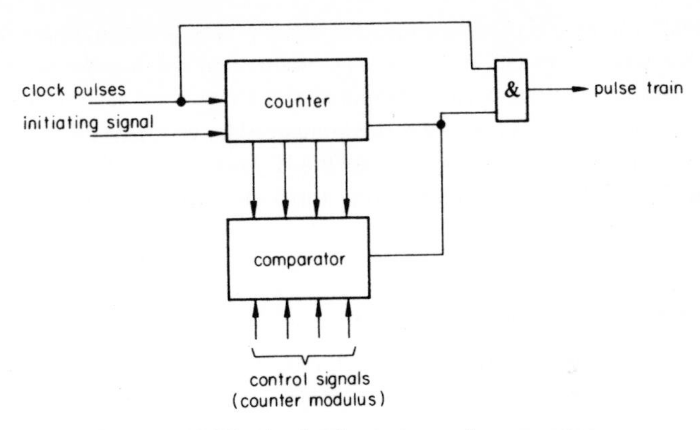

Fig. 13.13 The basis of a pulse train generator

The application of a single pulse to the **initiate** line causes a train of pulses to be generated by the circuit. So long as the value of the number stored in the counter is less than the value set by the control input lines, then the output from the comparator is logic '1'. This signal enables two functions to be carried out. Firstly, it allows clock pulses to continue to be counted. Secondly, it permits a train of pulses to appear at the output of the AND gate. When the number stored in the counter is equal to the value set by the control lines, the output from the comparator falls to zero. This signal stops the operation of the counter, and also prevents the further transmission of pulses through the AND gate.

The counter and comparator taken together form a **self-stopping variable modulus counter,** the modulus being set by the binary value of the signals on the control lines.

A circuit for the generation of *up to* fifteen pulses is shown in figure 13.14. A

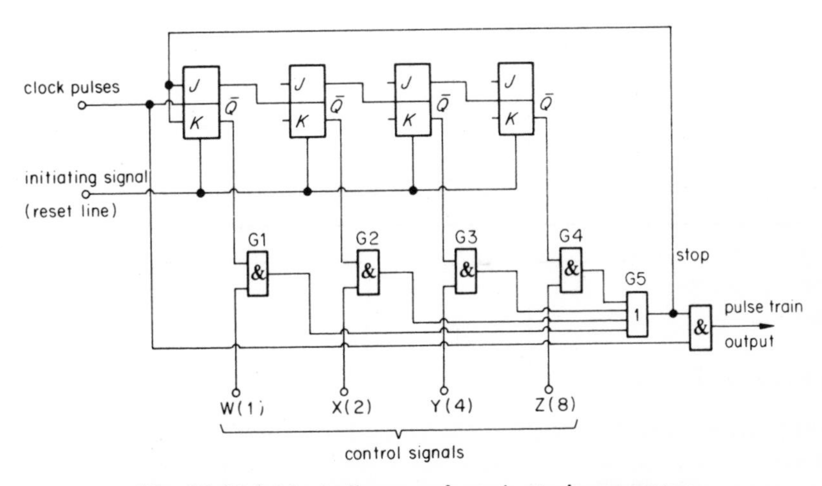

Fig. 13.14 A block diagram of a pulse-train generator

ripple-through counter is used to count the clock pulses, and the binary comparator is formed by gates G1 to G5, inclusive. The modulus of the counter is controlled by the signals applied to lines W to Z, the value of a logic '1' on each of the lines being shown in the figure. The pulse train is commenced by resetting all the flip–flops to zero, when all the $\bar{Q}$ outputs are reset to the '1' level. If $W = X = 1$, and $Y = Z = 0$, the output from G5 is '1' during the first three clock pulses, after which it is zero when counting ceases.

13.10 A trigger circuit using NAND gates

A trigger circuit is one which has two stable operating states, the output changing from one to the other when the input voltage reaches particular values set by the circuit. A popular application for a trigger circuit is as a voltage discriminator in which the output voltage from the trigger circuit changes abruptly from one level to another whenever the input voltage becomes either greater than or less than the threshold values.

Typical waveforms associated with this type of circuit are shown in figure 13.15. The input voltage level which causes the output voltage to change *when the input voltage is rising* is V_X, and *when the input voltage is falling* is V_Y. The difference

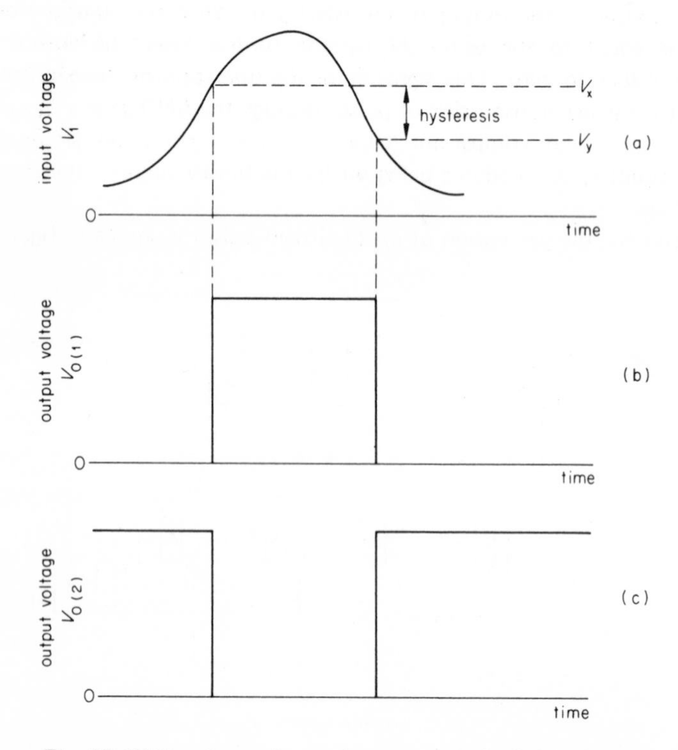

Fig. 13.15 Waveforms in a voltage-operated trigger circuit

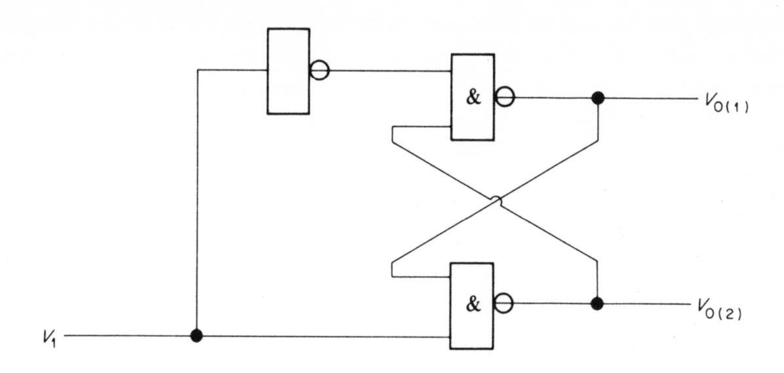

Fig. 13.16 A logic block diagram of a trigger circuit

between V_x and V_Y is known as the 'hysteresis' effect of the circuit. In practical circuits the hysteresis effect is advantageous since, in nearly all applications, the input signal has electrical noise superimposed upon it; in the absence of hysteresis, as the noise signal causes the net input voltage to change above and below the discrimination level, the output voltage would oscillate between zero and its maximum value. Typical output waveforms are shown in figures 13.15(b) and (c).

A simple trigger circuit using three gates is shown in figure 13.16. The threshold level V_Y is the voltage level at which the lower NAND gate turns ON, and voltage V_x is about 1 V above this value in DTL gates and in CMOS gates is some 3–4 V above V_Y.

Index

LIBRARY 46384
HARROW COLLEGE OF
HIGHER EDUCATION
NORTHWICK PARK
HARROW HA1 3TP.